NATURAL HISTORY
UNIVERSAL LIBRARY

西方博物学大系

主编：江晓原

NATURAL HISTORY OF BRITISH INSECTS

不列颠昆虫志

[英]爱德华·多诺万 著

华东师范大学出版社

图书在版编目（CIP）数据

不列颠昆虫志 = Natural history of British insects：
英文 /（英）爱德华·多诺万（Edward Donovan）著. ——
上海：华东师范大学出版社, 2018
（寰宇文献）
ISBN 978-7-5675-7988-0

Ⅰ.①不… Ⅱ.①爱… Ⅲ.①昆虫志-英国-英文
Ⅳ.①Q968.256.1

中国版本图书馆CIP数据核字(2018)第156662号

不列颠昆虫志

Natural history of British insects

（英）爱德华·多诺万（Edward Donovan）

特约策划　黄曙辉　徐　辰
责任编辑　庞　坚
特约编辑　许　倩
装帧设计　刘怡霖

出版发行　华东师范大学出版社
社　　址　上海市中山北路3663号　邮编 200062
网　　址　www.ecnupress.com.cn
电　　话　021-60821666　行政传真　021-62572105
客服电话　021-62865537
门市（邮购）电话　021-62869887
地　　址　上海市中山北路3663号华东师范大学校内先锋路口
网　　店　http://hdsdcbs.tmall.com/

印　刷　者　虎彩印艺股份有限公司
开　　本　787×1092　16开
印　　张　143.5
版　　次　2018年8月第1版
印　　次　2018年8月第1次
书　　号　ISBN 978-7-5675-7988-0
定　　价　2980.00元（精装全三册）

出　版　人　王　焰

总　目

《西方博物学大系》总序

江晓原

　　《西方博物学大系》收录博物学著作超过一百种，时间跨度为 15 世纪至 1919 年，作者分布于 16 个国家，写作语种有英语、法语、拉丁语、德语、弗莱芒语等，涉及对象包括植物、昆虫、软体动物、两栖动物、爬行动物、哺乳动物、鸟类和人类等，西方博物学史上的经典著作大备于此编。

中西方"博物"传统及观念之异同

　　今天中文里的"博物学"一词，学者们认为对应的英语词汇是 Natural History，考其本义，在中国传统文化中并无现成对应词汇。在中国传统文化中原有"博物"一词，与"自然史"当然并不精确相同，甚至还有着相当大的区别，但是在"搜集自然界的物品"这种最原始的意义上，两者确实也大有相通之处，故以"博物学"对译 Natural History 一词，大体仍属可取，而且已被广泛接受。

　　已故科学史前辈刘祖慰教授尝言：古代中国人处理知识，如开中药铺，有数十上百小抽屉，将百药分门别类放入其中，即心安矣。刘教授言此，其辞若有憾焉——认为中国人不致力于寻求世界"所以然之理"，故不如西方之分析传统优越。然而古代中国人这种处理知识的风格，正与西方的博物学相通。

　　与此相对，西方的分析传统致力于探求各种现象和物体之间的相互关系，试图以此解释宇宙运行的原因。自古希腊开始，西方哲人即孜孜不倦建构各种几何模型，欲用以说明宇宙如何运行，其中最典型的代表，即为托勒密（Ptolemy）的宇宙体系。

　　比较两者，差别即在于：古代中国人主要关心外部世界"如何"运行，而以希腊为源头的西方知识传统（西方并非没有别的知识传统，只是未能光大而已）更关心世界"为何"如此运行。在线性发展无限进步的科学主义观念体系中，我们习惯于认为"为何"是在解决了"如何"之后的更高境界，故西方的分析传统比中国的传统更高明。

然而考之古代实际情形，如此简单的优劣结论未必能够成立。例如以天文学言之，古代东西方世界天文学的终极问题是共同的：给定任意地点和时刻，计算出太阳、月亮和五大行星（七政）的位置。古代中国人虽不致力于建立几何模型去解释七政"为何"如此运行，但他们用抽象的周期叠加（古代巴比伦也使用类似方法），同样能在足够高的精度上计算并预报任意给定地点和时刻的七政位置。而通过持续观察天象变化以统计、收集各种天象周期，同样可视之为富有博物学色彩的活动。

还有一点需要注意：虽然我们已经接受了用"博物学"来对译 Natural History，但中国的博物传统，确实和西方的博物学有一个重大差别——即中国的博物传统是可以容纳怪力乱神的，而西方的博物学基本上没有怪力乱神的位置。

古代中国人的博物传统不限于"多识于鸟兽草木之名"。体现此种传统的典型著作，首推晋代张华《博物志》一书。书名"博物"，其义尽显。此书从内容到分类，无不充分体现它作为中国博物传统的代表资格。

《博物志》中内容，大致可分为五类：一、山川地理知识；二、奇禽异兽描述；三、古代神话材料；四、历史人物传说；五、神仙方伎故事。这五大类，完全符合中国文化中的博物传统，深合中国古代博物传统之旨。第一类，其中涉及宇宙学说，甚至还有"地动"思想，故为科学史家所重视。第二类，其中甚至出现了中国古代长期流传的"守宫砂"传说的早期文献：相传守宫砂点在处女胳膊上，永不褪色，只有性交之后才会自动消失。第三类，古代神话传说，其中甚至包括可猜想为现代"连体人"的记载。第四类，各种著名历史人物，比如三位著名刺客的传说，此三名刺客及所刺对象，历史上皆实有其人。第五类，包括各种古代方术传说，比如中国古代房中养生学说，房中术史上的传说人物之一"青牛道士封君达"等等。前两类与西方的博物学较为接近，但每一类都会带怪力乱神色彩。

"所有的科学不是物理学就是集邮"

在许多人心目中，画画花草图案，做做昆虫标本，拍拍植物照片，这类博物学活动，和精密的数理科学，比如天文学、物理学等等，那是无法同日而语的。博物学显得那么的初级、简单，甚至幼稚。这种观念，实际上是将"数理程度"作为唯一的标尺，用来衡量一切知识。但凡能够使用数学工具来描述的，或能够进行物理实验的，那就是"硬"科学。使用的数学工具越高深越复杂，似乎就越"硬"；物理实验设备越庞大，花费的金钱越多，似乎就越"高端"、越"先进"……

这样的观念，当然带着浓厚的"物理学沙文主义"色彩，在很多情况下是不正确的。而实际上，即使我们暂且同意上述"物理学沙文主义"的观念，博物学的"科学地位"也仍然可以保住。作为一个学天体物理专业出身，因而经常徜徉在"物理学沙文主义"幻影之下的人，我很乐意指出这样一个事实：现代天文学家们的研究工作中，仍然有绘制星图，编制星表，以及为此进行的巡天观测等等活动，这些活动和博物学家"寻花问柳"，绘制植物或昆虫图谱，本质上是完全一致的。

这里我们不妨重温物理学家卢瑟福（Ernest Rutherford）的金句："所有的科学不是物理学就是集邮（All science is either physics or stamp collecting）。"卢瑟福的这个金句堪称"物理学沙文主义"的极致，连天文学也没被他放在眼里。不过，按照中国传统的"博物"理念，集邮毫无疑问应该是博物学的一部分——尽管古代并没有邮票。卢瑟福的金句也可以从另一个角度来解读：既然在卢瑟福眼里天文学和博物学都只是"集邮"，那岂不就可以将博物学和天文学相提并论了？

如果我们摆脱了科学主义的语境，则西方模式的优越性将进一步被消解。例如，按照霍金（Stephen Hawking）在《大设计》（The Grand Design）中的意见，他所认同的是一种"依赖模型的实在论（model-dependent realism）"，即"不存在与图像或理论无关的实在性概念（There is no picture- or theory-independent concept of reality）"。在这样的认识中，我们以前所坚信的外部世界的客观性，已经不复存在。既然几何模型只不过是对外部世界图像的人为建构，则古代中国人干脆放弃这种建构直奔应用（毕竟在实际应用中我们只需要知道七政"如何"运行），又有何不可？

传说中的"神农尝百草"故事，也可以在类似意义下得到新的解读："尝百草"当然是富有博物学色彩的活动，神农通过这一活动，得知哪些草能够治病，哪些不能，然而在这个传说中，神农显然没有致力于解释"为何"某些草能够治病而另一些则不能，更不会去建立"模型"以说明之。

"帝国科学"的原罪

今日学者有倡言"博物学复兴"者，用意可有多种，诸如缓解压力、亲近自然、保护环境、绿色生活、可持续发展、科学主义解毒剂等等，皆属美善。编印《西方博物学大系》也是意欲为"博物学复兴"添一助力。

然而，对于这些博物学著作，有一点似乎从未见学者指出过，而鄙意以为，当我们披阅把玩欣赏这些著作时，意识到这一点是必须的。

这百余种著作的时间跨度为 15 世纪至 1919 年，注意这个时间跨度，正是西方列强"帝国科学"大行其道的时代。遥想当年，帝国的科学家们乘上帝国的军舰——达尔文在皇家海军"小猎犬号"上就是这样的场景之一，前往那些已经成为帝国的殖民地或还未成为殖民地的"未开化"的遥远地方，通常都是踌躇满志、充满优越感的。

作为一个典型的例子，英国学者法拉在（Patricia Fara）《性、植物学与帝国：林奈与班克斯》（*Sex, Botany and Empire, The Story of Carl Linnaeus and Joseph Banks*）一书中讲述了英国植物学家班克斯（Joseph Banks）的故事。1768 年 8 月 15 日，班克斯告别未婚妻，登上了澳大利亚军舰"奋进号"。此次"奋进号"的远航是受英国海军部和皇家学会资助，目的是前往南太平洋的塔希提岛（Tahiti，法属海外自治领，另一个常见的译名是"大溪地"）观测一次比较罕见的金星凌日。舰长库克（James Cook）是西方殖民史上最著名的舰长之一，多次远航探险，开拓海外殖民地。他还被认为是澳大利亚和夏威夷群岛的"发现"者，如今以他命名的群岛、海峡、山峰等不胜枚举。

当"奋进号"停靠塔希提岛时，班克斯一下就被当地美丽的土著女性迷昏了，他在她们的温柔乡里纵情狂欢，连库克舰长都看不下去了，"道德愤怒情绪偷偷溜进了他的日志当中，他发现自己根本不可能不去批评所见到的滥交行为"，而班克斯纵欲到了"连嫖妓都毫无激情"的地步——这是别人讽刺班克斯的说法，因为对于那时常年航行于茫茫大海上的男性来说，上岸嫖妓通常是一项能够唤起"激情"的活动。

而在"帝国科学"的宏大叙事中，科学家的私德是无关紧要的，人们关注的是科学家做出的科学发现。所以，尽管一面是班克斯在塔希提岛纵欲滥交，一面是他留在故乡的未婚妻正泪眼婆娑地"为远去的心上人绣织背心"，这样典型的"渣男"行径要是放在今天，非被互联网上的口水淹死不可，但是"班克斯很快从他们的分离之苦中走了出来，在外近三年，他活得倒十分滋润"。

法拉不无讽刺地指出了"帝国科学"的实质："班克斯接管了当地的女性和植物，而库克则保护了大英帝国在太平洋上的殖民地。"甚至对班克斯的植物学本身也调侃了一番："即使是植物学方面的科学术语也充满了性指涉。……这个体系主要依靠花朵之中雌雄生殖器官的数量来进行分类。"据说"要保护年轻妇女不受植物学教育的浸染，他们严令禁止各种各样的植物采集探险活动。"这简直就是将植物学看成一种"涉黄"的淫秽色情活动了。

在意识形态强烈影响着我们学术话语的时代，上面的故事通常是这样被描述的：库克舰长的"奋进号"军舰对殖民地和尚未成为殖民地的那些地方的所谓"访问"，其实是殖民者耀武扬威的侵略，搭载着达尔文的"小猎犬号"军舰也是同样行径；班克斯和当地女性的纵欲狂欢，当然是殖民者对土著妇女令人发指的蹂躏；即使是他采集当地植物标本的"科学考察"，也可以视为殖民者"窃取当地经济情报"的罪恶行为。

后来改革开放，上面那种意识形态话语被抛弃了，但似乎又走向了另一个极端，完全忘记或有意回避殖民者和帝国主义这个层面，只歌颂这些军舰上的科学家的伟大发现和成就，例如达尔文随着"小猎犬号"的航行，早已成为一曲祥和优美的科学颂歌。

其实达尔文也未能免俗，他在远航中也乐意与土著女性打打交道，当然他没有像班克斯那样滥情纵欲。在达尔文为"小猎犬号"远航写的《环球游记》中，我们读到："回程途中我们遇到一群黑人姑娘在聚会，……我们笑着看了很久，还给了她们一些钱，这着实令她们欣喜一番，拿着钱尖声大笑起来，很远还能听到那愉悦的笑声。"

有趣的是，在班克斯在塔希提岛纵欲六十多年后，达尔文随着"小猎犬号"也来到了塔希提岛，岛上的土著女性同样引起了达尔文的注意，在《环球游记》中他写道："我对这里妇女的外貌感到有些失望，然而她们却很爱美，把一朵白花或者红花戴在脑后的髮髻上……"接着他以居高临下的笔调描述了当地女性的几种发饰。

用今天的眼光来看，这些在别的民族土地上采集植物动物标本、测量地质水文数据等等的"科学考察"行为，有没有合法性问题？有没有侵犯主权的问题？这些行为得到当地人的同意了吗？当地人知道这些行为的性质和意义吗？他们有知情权吗？……这些问题，在今天的国际交往中，确实都是存在的。

也许有人会为这些帝国科学家辩解说：那时当地土著尚在未开化或半开化状态中，他们哪有"国家主权"的意识啊？他们也没有制止帝国科学家的考察活动啊？但是，这样的辩解是无法成立的。

姑不论当地土著当时究竟有没有试图制止帝国科学家的"科学考察"行为，现在早已不得而知，只要殖民者没有记录下来，我们通常就无法知道。况且殖民者有军舰有枪炮，土著就是想制止也无能为力。正如法拉所描述的："在几个塔希提人被杀之后，一套行之有效的易货贸易体制建立了起来。"

即使土著因为无知而没有制止帝国科学家的"科学考察"行为，这事也很像一个成年人闯进别人的家，难道因为那家只有不懂事的小孩子，闯入者就可以随便打探那家的隐私、拿走那家的东西、甚至将那家的房屋土地据为己有吗？事实上，很多情况下殖民者就是这样干的。所以，所谓的"帝国科学"，其实是有着原罪的。

如果沿用上述比喻，现在的局面是，家家户户都不会只有不懂事的孩子了，所以任何外来者要想进行"科学探索"，他也得和这家主人达成共识，得到这家主人的允许才能够进行。即使这种共识的达成依赖于利益的交换，至少也不能单方面强加于人。

博物学在今日中国

博物学在今日中国之复兴，北京大学刘华杰教授提倡之功殊不可没。自刘教授大力提倡之后，各界人士纷纷跟进，仿佛昔日蔡锷在云南起兵反袁之"滇黔首义，薄海同钦，一檄遥传，景从恐后"光景，这当然是和博物学本身特点密切相关的。

无论在西方还是在中国，无论在过去还是在当下，为何博物学在它繁荣时尚的阶段，就会应者云集？深究起来，恐怕和博物学本身的特点有关。博物学没有复杂的理论结构，它的专业训练也相对容易，至少没有天文学、物理学那样的数理"门槛"，所以和一些数理学科相比，博物学可以有更多的自学成才者。这次编印的《西方博物学大系》，卷帙浩繁，蔚为大观，同样说明了这一点。

最后，还有一点明显的差别必须在此处强调指出：用刘华杰教授喜欢的术语来说，《西方博物学大系》所收入的百余种著作，绝大部分属于"一阶"性质的工作，即直接对博物学作出了贡献的著作。事实上，这也是它们被收入《西方博物学大系》的主要理由之一。而在中国国内目前已经相当热的博物学时尚潮流中，绝大部分已经出版的书籍，不是属于"二阶"性质（比如介绍西方的博物学成就），就是文学性的吟风咏月野草闲花。

要寻找中国当代学者在博物学方面的"一阶"著作，如果有之，以笔者之孤陋寡闻，唯有刘华杰教授的《檀岛花事——夏威夷植物日记》三卷，可以当之。这是刘教授在夏威夷群岛实地考察当地植物的成果，不仅属于直接对博物学作出贡献之作，而且至少在形式上将昔日"帝国科学"的逻辑反其道而用之，岂不快哉！

2018 年 6 月 5 日
于上海交通大学
科学史与科学文化研究院

　　《不列颠昆虫志》是爱德华·多诺万（Edward Donovan，1768—1837）的一部博物学著作。多诺万是英国博物学家，生于爱尔兰的科克，曾是伦敦林奈学会会员。在那个博物学家和收藏家纷纷开设个人博物馆的时代，1807年他也在伦敦设立了博物学研究所，展示数百件动物标本与植物标本。他本人并不去海外采集标本，而是运用自己良好的社会关系，委托包括约瑟夫·班克斯和詹姆斯·库克在内的诸多探险家为自己工作，因而得以聚集起大量罕见标本。对这些标本的研究帮助他出版了诸多负有盛名的博物学著作，如《不列颠珍稀鸟类志》《中国昆虫志》《印度昆虫志》。多诺万亲自制作铜版并调色，使书中的画色彩尽可能鲜艳逼真，加上展示了一批与他有特殊关系的相关人士搞到的珍稀标本，因此这些著作一经出版其声名就居于当时同类作品的最前列。

　　《不列颠昆虫志》初版刊行于1792年，此时作者的事业正值盛期。全书凡十六卷共3520页，附有约500幅根据实体标本测量、描绘的精美彩图。和之前的《不列颠贝类志》一样，作者在设计绘制这些插画时，不仅进行单体展示，而且根据分类并通过巧妙的构图，在每幅画中展现各类昆虫的实际大小比例、同一昆虫的生长变态各阶段，并往往将它们画进生活环境中。这样的创作，无扎实的昆虫学知识绝难胜任。

　　今据原版影印。

THE
NATURAL HISTORY
OF
BRITISH INSECTS;

EXPLAINING THEM

IN THEIR SEVERAL STATES,

WITH THE PERIODS OF THEIR TRANSFORMATIONS,
THEIR FOOD, ŒCONOMY, &c.

TOGETHER WITH THE

HISTORY OF SUCH MINUTE INSECTS

AS REQUIRE INVESTIGATION BY THE MICROSCOPE.

THE WHOLE ILLUSTRATED BY

COLOURED FIGURES,

DESIGNED AND EXECUTED FROM LIVING SPECIMENS.

BY E. DONOVAN.

VOL. I.

LONDON:

PRINTED FOR THE AUTHOR,

And for F. and C. RIVINGTON, Nº 62, ST. PAUL'S CHURCH-YARD.

MDCCXCII.

A

SLIGHT SKETCH

OF THE

ANIMAL SYSTEM.

LINNÆUS divided the Animal Syſtem into ſix claſſes.

Claſs I. MAMMALIA. Suckle their young.
   II. AVES. (Birds) covered with feathers.
   III. AMPHIBIA. Lungs arbitrary.
   IV. PISCES. (Fiſhes) breath by gills not arbitrarily.
   V. INSECTA. (Inſects) two antennæ, or feelers *.
   VI. VERMES. No head.

We therefore ſee that Inſects compoſe the fifth Claſs in the Syſtem, and are divided into ſeven Orders.

Order I. COLEOPTERA. Wings two, covered by two ſhells divided by a longitudinal ſuture.
   II. HEMIPTERA. Shells or covers of the wings, ſomewhat ſoft, and incumbent on each other.
   III. LEPIDOPTERA. Wings four, imbricated with minute ſcales.

* Thoſe feelers are the two horns that are affixed to the head.

B 2

IV. NEU

-3-

PLATE I.

IV. NEUROPTERA. *Wings four, naked, transparent, reticulated, with veins or nerves. Tail without sting.*

V. HYMENOPTERA. *Wings four. Membraneous; tail of the female armed with a sting.*

VI. DIPTERA. *Wings two.*

VII. APTERA. *No wings.*

TRANSFORMATIONS OF INSECTS.

Many of our readers are no doubt acquainted with the singular transformations Insects undergo, but we trust those will pardon a digression which may be useful to those who have not that knowledge; and without premising farther we proceed to inform them, that Insects in general undergo a material change in their form at stated periods of their lives; there are some, though few, which burst forth from the egg perfectly formed, as *Spiders*, &c. but the greater part exist in four several states: the first that of the egg, whence the Larva, or Caterpillar is produced; it is at first very minute, but in this state it feeds, some kinds on one or two plants only, others promiscuously on many, they therefore continue to increase in size, moulting several times the outer skin, until the destined period of their dormant state approaches; they then spin a web more or less strong according to the species, and are converted into the aurelia, or chrysalis; and lastly they burst forth in due season with all their accomplishments perfect. It is under this form they propagate a future race, and themselves perish, as they rarely survive the inclemencies of the winter.

The antient naturalists held suppositions very imperfect and erroneous relative to those transformations, but *Malpighi* and *Swammerdam* proved by many accurate examinations clearly, that those changes were not suddenly effected, but gradual; and that under the form of the Caterpillar they could distinguish the future changes the Insect would undergo.

PLATE

PLATE I.

PHALÆNA PAVONIA,

EMPEROR MOTH.

LEPIDOPTERA.

Infects of the LEPIDOPTERA ORDER are divided into three *Genera*, PAPILIO, SPHINX, and PHALÆNA, *Butterflies*, *Hawk Moths*, and *Moths*. The characters of the two former hereafter : those of the Phalæna are

GENERIC CHARACTER.

The antennæ fetaceous, decreafing in fize from the bafe to the apex. The wings, when at reft, are generally deflected. They fly in the night.

SPECIFIC CHARACTER.

Antennæ feathered. No trunk. Wings expanded, horizontal, rounded, entire, with a large eye in the center of each ; the firft red-brown waved ; the fecond orange. The antennæ of the male are broader, and the wings of the female larger, waved with black and white and bordered with yellow. Caterpillar green or yellow, fpinous, on thorns and brambles. Length of the moth one inch.—*Berken. Out.*

The conformity and likenefs which prevails between the male and female throughout the greater part of the animal fyftem, cannot however in infects be implicitly depended on; the difference in many is fuch as even to miflead fome very accurate entomologifts, the illuftrious Linnæus not excepted. In this fpecies it is not fo great as

in

PLATE I.   7

in many, but such as entitles it to a figure in a future plate; the want of room determining us to defer it for the present. Our figure is that of the male.

Albin, (*Plate* 25, *Subject* 37,) has given a figure of the male and female on the same plate, and describes a male to have changed to the aurelia state as in our plate represented *July* 16, and *March* 18 following to have produced the Fly. But the time of their appearance depends on the proportion of heat and cold; what the author mentions was preserved from the severity of winter, in a warm room; the usual time to find them in the caterpillar state is August, and in April the fly.

The singular provision which nature makes for the protection of this Fly deserves particular notice; when the time of its continuation in the caterpillar state is expired, like all others, it refuses to eat; it then, by much labour, forms a kind of bag or purse, of a very tough substance; this it fixes against the trunks of trees, &c. by a number of hairs or filaments, which remain on the external surface. It lines the outer case by one of a finer texture, the top of which is closed by several bristles that unite in the center, exactly representing a cap, and excludes almost the possibility of its receiving an injury during this defenceless state. In this bag it passes to the aurelia, and remains until the birth of the perfect insect.—Our figure represents the chrysalis or aurelia in the bag; part appears torn away to exhibit its situation therein.

Were we to unite the several accounts of authors respecting its food it would appear a general feeder; it will however live on the rose, the elm, and the willow; and on thorns and brambles particularly.

PLATE

PLATE II.

FIG. I.

MONOCULUS QUADRICORNIS.

APTERA.

Apterous infects are diftinguifhed chiefly by having no wings in either male or female.

GENERIC CHARACTER.

The feet are formed for fwimming. The body is covered with a cruftaceous cafe or fhell. The eyes fixed in the fhell very near each other.

SPECIFIC CHARACTER.

Grey brown. One eye. Antennæ four. Body diminifhes towards the tail, which is long and bifid, with three or four ftrong hairs on each fide. A bag of eggs on each fide of the tail. Length half a line.—*Berk. Out.*

Although this infect may have been noticed by many fwimming, or rather darting fwiftly in various directions in water; its minutenefs is fuch, that the moft attentive could never have comprehended precifely its component parts; but the microfcope difcovers it to be an animal of fuch fingular formation as highly to deferve the attention of the naturalift. It is covered by a firm cruftaceous fubftance, divided into annulations, and armed in feveral parts with fpines and briftles; notwithftanding

9

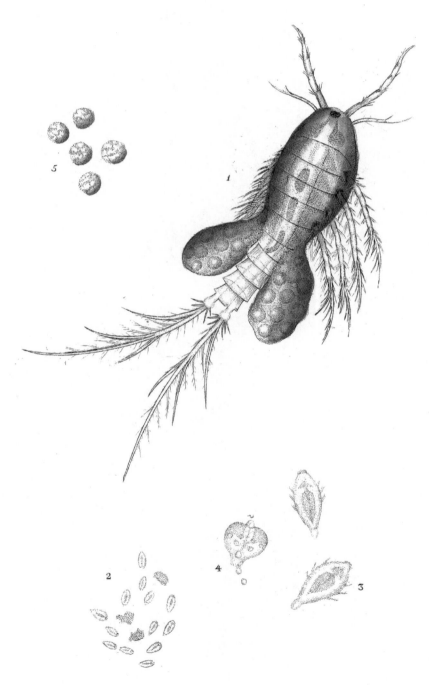

PLATE II. 9

withftanding which, this fhell is fo tranfparent that the whole motion of the inteftines is very vifible by a good magnifier.

It muft be granted that, but for the microfcope, the wonders of the minute creation would be to us entirely unknown, our ideas could never fuppofe the exiftence of thofe animated forms which occupy the immeafurable fpace between an apparent atom and nothing. The myriads of animations, thoufands of times fmaller than a mite, muft evade our cognizance, and be an actual conviction of their non-exiftence.

But with all the utility that the microfcope can boaft, no inftrument is fo likely to miflead the moft accurate obferver, particularly if not in the habit of ufing it; the variations of light, the powers of magnification, or the damage the glaffes may meet with by accident, fuch as requires every one to examine with the greateft care; one degree of light may bring an object to view, whilft another may entirely blend it with the fluid it exifts in; or one glafs may difcover fpines on an object, another glafs might have reprefented perfectly fmooth; it is therefore neceffary to begin with a fmall power, in proportion to the fize of the object, and to proceed to deeper magnifiers after.

There is fome difference in our figure and thofe either of *Barbut*, or of *Baker*, which appears chiefly from our ufing a fingle lens nearly of the deepeft power convenient to ufe. Our glaffes were the 20th and 30th of an inch focus.

We very attentively examined the eyes, and found, not one, but two, placed near each other, on a fcale or plate of a black colour; hence arifes the appearance of a fingle eye by a fmall magnifying power.

The tail prefents a forked appearance by a deep power, and the eggs are contained in two bags, one on each fide the tail. The colour varies probably in proportion to the nature of its food, to pale green, more or lefs of a red, or of a grey brown colour.

Fɪɢ.

F I G.   II.

This minute animalcula is frequent in ftagnant water, or in infufions of vegetables, and is one fpecies of thofe whofe exiftence can only be difcovered by a good microfcope. It is very difficult, confidering the power thofe creatures have to diftort their true form at pleafure, to fix their diftinguifhing character : therefore where the definition appears dubious, we prefer being filent rather than hazard an error.

FIG. 2. Reprefents them (*magnified*) as they fometimes feem to follow the leader in herds ; but perhaps it is only the fcent of the prey that induces each to follow the foremoft, as they frequently fwim or whirl in the water feparately, with great fwiftnefs, devouring the fmaller kinds of animalculæ.

FIG. 3. Two, magnified by a deep power, when they appear to have feet or fins.

FIG. 4. Shews the ftrange form it affumes to depofit its eggs.

FIG. 5. The Eggs deeper magnified.

P L A T E

PLATE III.

PHALÆNA BUCEPHALA,

BUFF-TIP MOTH.

LEPIDOPTERA.

GENERIC CHARACTER.

Antennæ taper from the bafe to the apex, and are fetaceous.
Wings in general defle&ed when at reft. Fly by night. No Trunk.
Wings reverfed, i. e. firft Wings horizontal and fecond ere&.

SPECIFIC CHARACTER.

Antennæ feathered. Firft Wings grey, with two double tranfverfe
brown waves, and a large yellowifh brown fpot at the extreme angle.
Second Wings plain, light yellow, length fcarce one inch. Caterpillar hairy, yellow with black fpots. On Oaks, Afh, &c.—*Berkenhout.*

The delicate affemblage of beautiful down which cloath the upper
wings of the Buff-tip Moth is its chief recommendation; the hiftory
affords but little for obfervation, it is hatched from the egg in *Auguft*,
and in *June* following the fly is perfe&.

Its beauty avails not the race of birds who purfue them from
neceffity, or from an innate defire of cruelty and devaftation; and
whilft

whilſt happy in its apparent ſecurity, ranging the plain to experience
the pleaſures of liberty, to banquet in the nectareous profuſion of the
vegetable kingdom, or catch the dew-drop from the humid air, to
inſpirit and refreſh his parched ſyſtem from the mid-day heat, he becomes a dupe to his happineſs, his pleaſures at once fully, and he falls
an unreſiſting victim into the devouring jaws of death.

PLATE IV.

PHALÆNA GROSSULARIATA.

Magpye, or Currant-Moth.

Generic Character.

The antennæ fetaceous, decreafing in fize from the bafe to the point. The Wings, when at reft, generally deflected. Fly by night.

Antennæ taper, like briftles.

Specific Character.

Antennæ and Legs black. Body yellow, with black fpots. Wings white, with many black patches, and a tranfverfe yellow wave on the firft pair. Caterpillar white, with black fpots ·on the Back; Belly yellow. *Berk. Out.*

The Magpye-Moth is one of the *geometræ*; and feeds on Goofberry and Currant-bufhes, as it's name indicates. The Caterpillar is found in *May*; and in *July*, the Fly.

The Caterpillar, previous to its change to the Chryfalis ftate, fpins a web of a very flight and delicate texture, by which it is fufpended horizontally againft the branches of trees, &c. as in our Plate re-prefented.

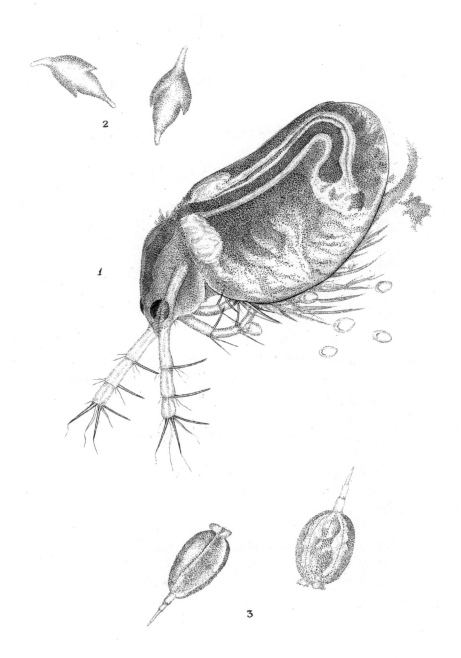

5

2

1

3

PLATE V.

FIG. I.

MONOCULUS CONCHACEUS.

APTERA.

Without wings.

GENERIC CHARACTER.

Body covered with a cruft or fhell. Feet made for fwimming.

SPECIFIC CHARACTER.

Inclofed in a bivalve, afh-coloured fhell, from the aperture of which it puts forth a number of capillary antennæ, which it retracts when taken out of the water.

To acquire a proper knowledge of the formation of this minute infect, it is neceffary to ufe a microfcope with a glafs ftage for objects, or rather fuch as admit of nicely adjufting a talc as occafion may require: the infect to be taken from the water with a camel-hair pencil, and carefully placed on the talc; after which it may be examined by a magnifier of $\frac{1}{6}$ of an inch focus; but in proceeding to a deeper power, let the talc be turned the upper furface with the infect in the drop of the fluid from the lens, and thereby the lens may approach the object to its proper focus; without this caution the lens would be frequently immerged in the water, and entirely obftruct the fight.

In the adult ftate, the opacity of the bivalve fhell, its external covering, fo entirely obfcures the internal motion, that nothing, except the filaments it throws from the aperture or opening, is vifible by the microfcope.

It breaks from the egg perfectly formed, but very minute and tranfparent; this is therefore the beft time to difcover its ftructure, and from one in this ftate we have taken our figure.

5

By

By the antennæ it directs its courſe, as does the *Monoculus Quadri-cornis* ; and like it alſo it hath two eyes fixed in the ſhell, but it can com-pletely envelop its head in its bivalve covering ; its mouth is beneath, but the numerous filaments it darts forth, cauſes ſuch a violent motion in the water, that the minuter inſects are unreſiſtingly drawn between them, and forced to the mouth.

The motion of its lungs is very viſible, as are alſo the veſſels rami-fying thence. Its food is carried to, and digeſted in the deep-coloured tube, or inteſtine, and the refuſe is diſcharged by a ſudden jerk from the extremity of the tube, or anus.

Thus it exiſts, a life of rapine and deſtruction, enjoyed at the ex-pence of the lives of thouſands ; and as the objects of its ravenous diſ-poſition are defenceleſs, ſo are they the ſport of their conqueror : the few moments of intermiſſion its craving appetite grants them, is occu-pied equally in the ſpoil, firſt preſſing them to death, and then toſſing them undevoured into the fluid.

But ſhould a more powerful inſect oppoſe him, he immediately con-tracts his parts, and nothing more than the external covering is open to his antagoniſt's violence, and he will ſooner die ignobly than offer the leaſt oppoſition.

—————

FIG. II.

This animalcule is very minute, and appears like a fine membrane without inteſtines before the microſcope ; from the appearance of its winged ſides, it is ſuppoſed to reſemble a bird. It is called *Burſaria Hirundinella.*

—————

FIG. III.

The back and ſide view of an animalcule found in ditch-water on duck-weed, very pellucid, and ſingularly marked in the inteſtines ; tail moveable, and thereby it directs its courſe.

PLATE VI.

SPHINX FILIPENDULÆ.

BURNET MOTH.

GENERIC CHARACTER.

Sphinx, Antennæ thickeſt in the middle. Wings, when at reſt, deflexed. Fly ſlow, morning and evening only.

SPECIFIC CHARACTER.

Antennæ, Legs, and Body black. Second Wings red, with a greeniſh border. Firſt Wings bluiſh green, with ſix red ſpots, in pairs, length eight lines. Caterpillar yellow, with black ſpots. *Berk. Out.*

The female has but five red ſpots on the upper Wing, the two ſpots at their baſe being placed ſo near each other as only to form one large ſpot.

It feeds on the *Geniſta Anglica,* needle furze ; on the *Ulex Europœus,* common furze ; and on the filipendula.

The Caterpillars of moſt of the inſects of this genus are armed with a ſpine or horn above the anus, in which particular this differs. It is in the Caterpillar ſtate in *May,* and *June,* and in *July* the Sphinx.

PLATE VII.

CHRYSIS IGNITA.

HYMENOPTERA.

GENERIC CHARACTER.

The abdomen hath three annulations exclufive of the anus, the antennæ hath twelve articulations, exclufive of the firft joint which is longer than the reft. The body fhines like polifhed metal. A kind of collar is very diftinct in this *genus*. The anus is dentated, having one, two, or more teeth.

SPECIFIC CHARACTER.

The antennæ are black, the thorax a fine mazarine blue, having in fome portions a greenifh caft, the abdomen a fine gold colour with fhades of crimfon and yellow green; the anus hath four teeth or denticulations.—*Harris Inf.*

Exotic Infects, or at leaft thofe of the Eaft, and Weft Indies, for the effulgence, and beauty, of their colouring in general, claim a fuperiority over the natives of this climate; but the appearance of this Chryfis before the fpeculum of an opake microfcope, may vie with many of the moft favourite foreigners hitherto difcovered: the richnefs

D

of

of changeable colours blending into each other, according to the variations of the light reflected on the furface, is fuch that we freely con-
fefs our inability, or even the inability of art, to equal, though we
truft our figure will give fome idea of the delightful appearance of the
original.

The Fly of the natural fize is given on the fore ground, the magnified figure above.

It is found againft decayed trees or walls, in the hotteft fun-fhine
of Summer.

P L A T E

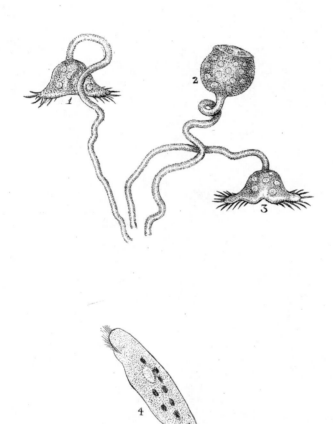

PLATE VIII.

VORTICELLA LUNARIS.

GENERIC CHARACTER.

A worm capable of contracting or extending itself, naked, with rotatory cilia.

SPECIFIC CHARACTER.

Simple, hemifpherical, with a twifted pedicle.

" The fmall head of this animalculum is crater-form, the margin
" of the orifice protuberant, ciliated on both fides, the hairs undula-
" ting, the pedicle eight or ten times the length of the body. As
" often as the mouth is opened, the pedicle extends itfelf; when it is
" fhut, this is twifted up fpirally, and their motions are often reite-
" rated in a fhort fpace.

" FIG. 1. the head, expanded.——FIG. 2. when fhut.——FIG. 3.
" the undulated edge."

Adams's Effays on the Microfcope.

FIG. 4. found in infufions of hay; and is called *Trichoda Uvula.*

D 2   PLATE

PLATE IX.

PHALÆNA EVONYMELLA.

SMALL ERMINE MOTH.

LEPIDOPTERA.

GENERIC CHARACTER.

Antennæ taper like briftles.

SPECIFIC CHARACTER.

Firft wings filver-white, with fifty fmall black fpots in rows. Second wings lead-colour.

Phalæna Evonymella feeds on the white-thorn, black-thorn, and on fruit-trees; in May the caterpillars are hatched, and as they live in focieties of hundreds, or even thoufands, by their united induftry they fpin a web fpacious enough to contain the family, and therein they affume their feveral forms; early in June they become chryfalides, and in about fourteen days the Flies are perfect.

The caterpillars of the *Pha. Padella* and *Evonymella* are ever found in the fame fociety, and many circumftances may be advanced

4   to

PLATE IX.

23

to prove them either varieties of each other, or difference of fex only, although Linnæus confidered them as diftinct fpecies. They differ in colour, the caterpillars of one being light yellow brown, the other black, and the upper wings of the *Evonymella* are lefs of a lead colour than thofe of the *Padella*.

To gain information on this fubject, we, this feafon, put the eggs of feveral females into different glaffes; the eggs of each female produced both kinds of caterpillars, they became chryfalides, and a number of each fort of the Flies came forth.

PLATE

PLATE X.

PHALÆNA CHRYSORRHŒA.

YELLOW TAIL MOTH.

LEPIDOPTERA.

GENERIC CHARACTER.

No trunk. Wings depreffed, deflexed. Back fmooth.

SPECIFIC CHARACTER.

Antennæ feathered. Entirely white, except the extremity of the abdomen, which is yellow. Caterpillar black and red, hairy.— *Berk. Out.*

Linnæus in the *Syftema Naturæ*, has confounded the *Yellow Tail*, with the *Brown Tail*, *Moth*, nor was it generally confidered as an error till fome time after ; but the immenfe increafe of the caterpillars of the Brown Tail Moth in the year 1780, afforded an opportunity of determining them to be diftinct fpecies.

Though foreign to our purpofe, and properly under the hiftory of the Brown Tail Moth, we cannot pafs over fuch remarkable circumftances as attended the uncommon increafe of this fpecies in the above winter.
The

PLATE X. 25

The fears of the public muſt have been great indeed, when prayers were offered to avert the famine ſuppoſed to be threatened by the appearance of thoſe inſects in the ſtate of the caterpillar.

In July the Caterpillar is found feeding on the white-thorn, ſallow, apple-trees, and on fruit-trees in general, about the latter end of the ſame month it ſpins a web of a tough texture againſt the branches of trees, &c. becomes an aurelia, and in Auguſt the Fly comes forth.

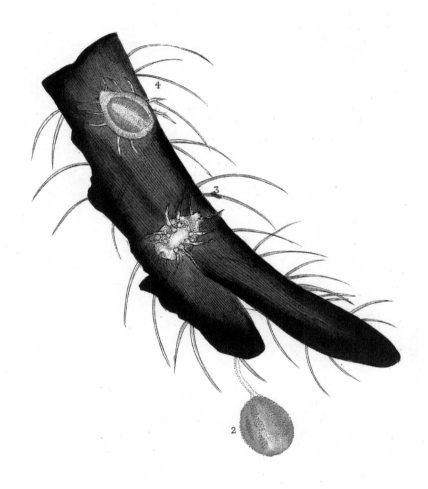

11

PLATE XI.

ACARUS COLEOPTRATORUM.

BEETLE-TICK.

APTERA.

GENERIC CHARACTER.

Legs eight. Eyes two, lateral. Tentaculæ two, jointed.

SPECIFIC CHARACTER.

Tawny. Anus whitiſh.

This Tick is one of thoſe deteſtable race of animals whoſe minuteneſs ſecures it from danger, while it draws nutriment from the blood, and frequently from the vitals of larger inſects. Every animal is tormented by thoſe cruel and blood-thirſty beings, varying in ſize, in ſhape, and in colour, but whether they be diſtinguiſhed by the name of lice, bugs, fleas, or mites, they fall under one point of view, when conſidered as a peſt to the ſocieties of other animals.

Beetles are in general infeſted and ſeverely injured by thoſe vermin. I found about a month ſince one of the *Scarabæus Stercorarius, Com-*

E

mon

mon Dor, or *Clock*, almost devoured alive by them; little except his shell remaining; yet, in this state it lived several days. There were a number of small brown bags affixed by pedicles to its breast, thighs, and even feet; the microscope discovered those to contain each an embryo, and the pedicle, no doubt, answered the part of an umbilical chord, to extract nourishment from the living creature. I perceived on further inspection their base penetrated the shell, or entered the apertures.

FIG. 1. Natural size of the Tick and Embryo.

FIG. 2. The upper side, and FIG. 3. under side, magnified.

> GRADUAL, from these what numerous kinds descend,
> Evading even the microscopic eye!
> All Nature swarms with life; one wond'rous mass
> Of Animals or Atoms organized,
> Waiting the vital breath, when PARENT HEAVEN
> Shall bid his Spirit blow. —— —— —— ——
> —— —— —— —— —— ——
> —— —— —— —— These, conceal'd
> By the kind art of forming HEAVEN, escape
> The grosser eye of man: for, if the worlds
> In worlds inclos'd, should on his senses burst,
> From cates ambrosial, and the nectar'd bowl
> He would abhorrent turn; and in dead night
> When silence sleeps o'er all, be stunn'd with noise.
>
> THOMSON's SEASONS.

PLATE

PLATE XII.

CICINDELA CAMPESTRIS.

SPARKLER.

COLEOPTERA.

GENERIC CHARACTER.

Antennæ taper. Jaws prominent, denticulated. Eyes prominent, Thorax margined. Five joints in each foot.

SPECIFIC CHARACTER.

Above green-gold. Beneath copper tinged. Eyes large. Thorax angular and narrower than the head. Six fpots on each fhell. An oval fubftance at the bafe of each thigh. Legs long and flender.

————————————

This beautiful infect varies fomething in fize and colour, the fpots on the elytra are generally white, but are often found with fpots of yellow; they fly or run quick, are carnivorous, and live in dry fandy places. In the fpring its larva is found, which refembles a long, foft, whitifh worm, with fix legs and a brown fcaly head; it perforates

9   the

the fand perpendicularly, and refts near the furface to enfnare fmaller infects.

It is very difficult, if at all poffible, to breed thofe infects and obferve their metamorphofes; we have tried various methods, but have not yet been fo fortunate as to fucceed.

PLATE XIII.

LUCANUS CERVUS.

STAG BEETLE.

COLEOPTERA.

GENERIC CHARACTER.

Antennæ clavated, compreſſed, pectinato-fiſſile. Maxillæ extended ſo as to reſemble horns. Five joints in each foot.

SPECIFIC CHARACTER.

Head and Thorax black. Shells dark brown. Horns reſembling thoſe of a Stag, forked at the end, a ſmall branch near the middle on the inſide, moveable. Shells plain.

———————————————

The Stag-Beetle is the largeſt coleopterous inſect we poſſeſs, but its ſize is inſignificant, when compared with thoſe of the ſame kind that inhabit hot countries or woodlands, as inſtanced in the *Scarabæus Hercules*, &c.

F

It

It is fufficiently diftinguifhed in this country by the moveable maxil-læ, or jaws, that projeᵭt from the head; they are of a dark red co-lour, and though brighter in fome fpecimens, are rarely of the beau-tiful coral appearance *Barbut* and other authors have defcribed.

Coleopterous infeᵭts in general are endowed with amazing ftrength, and their arms are equally ferviceable for the affault or defence. The antlers of this Beetle are carefully to be avoided by fuch as attempt to deprive it of liberty; with them it ftrips off the bark of oak trees, and attaches itfelf firmly to the trunk, thence extraᵭting the liquor that oozes with its tongue.

They are plentiful in June and July, in Kent and Effex, and in many other parts of Britain.

The females are known by their maxillæ being much fhorter than thofe of the males; they depofit their eggs under the bark of old trees, either oak or afh, and the food of the larvæ, or grubs, is the internal fubftance of the trunk, firft reduced to a fine powder; they undergo transformation in this cell, and force a paffage through the bark when perfeᵭt beetles.

PLATE

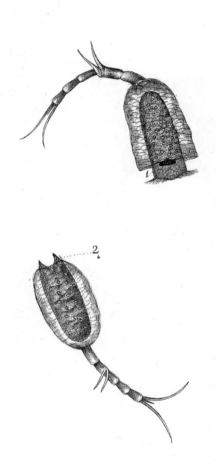

PLATE XIV.

TRICHODA POCILLUM.

TRICHODA.

An invisible, pellucid, hairy worm.

SPECIFIC CHARACTER.

Oblong trichoda, the fore-part truncated and hairy, the tail articulated, and divided into two bristles.

———————

This invisible animalculum is common in marshy places, particularly in the swamps near the banks of the river Thames.

When magnified, the body is pellucid, and appears as two separate bodies, one enclosing the other; the interior part is filled with molecules, and the exterior is membranaceous: they are capable of extension or dilation, and of folding in various directions. At the extremity of the interior part is a muscular orbicular membrane, which is opened or shut at pleasure, and forms the mouth.

FIG. 1. The interior part protruded with the mouth open.

FIG. 2. The jaws shut.

F 2   PLATE

PLATE XV.

PHALÆNA CAJA.

GREAT TYGER MOTH.

LEPIDOPTERA.

GENERIC CHARACTER.

Antennæ taper from the base. No trunk. Wings depreffed, deflexed. Back fmooth.

SPECIFIC CHARACTER.

Antennæ feathered. Firft wings whitifh, with large irregular dark brown fpots. Abdomen and fecond wings orange, with black fpots.

———————

The fuperior wings in fome of this fpecies have fmaller brown fpots, and more of the cream colour; in others the fpots are larger, and frequently two are united to form one. The inferior wings alfo admit of equal variety; the fpots near the thorax are often united, and the fmall black ftripes on the back are fewer in the prefent fpecimen than are common to the Moth.

The

The caterpillars feed on lettuces, or nettles. When he is appre-
henfive of danger, he rolls himfelf up like a hedge-hog. He becomes a chryfalis in May; and the latter end of June, or early in
July, it produces the Moth.

PLATE

PLATE XVI.

PHALÆNA ANTIQUA.

WHITE SPOT TUSSOCK MOTH,
OR
VAPOURER.

LEPIDOPTERA.

GENERIC CHARACTER.

Antennæ taper from the base. No trunk. Wings depressed. Back hairy.

SPECIFIC CHARACTER.

Antennæ feathered. First wings cloudy, orange, waved and spotted with brown, and a white spot on the posterior angle. Female without wings.

The female Vapourer Moth at first sight perfectly resembles an apterous insect; but on inspection, very small wings are seen at the extremity of the Thorax, and the antennæ determine it to be a phalæna. It creeps in a sluggish manner, and lays an abundance of eggs.

FIG. 1. the Female. FIG. 2. the Male.

The

The Caterpillars feed on white thorn, and on trees in general. It has been known to live on the deadly night-fhade, and other poifonous plants. The Caterpillars are found in July, and the Moth in September.

17

2　1

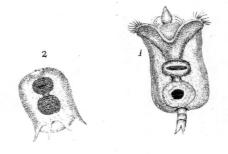

3

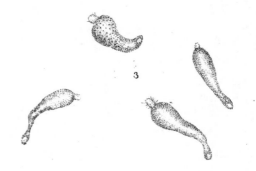

PLATE XVII.

VORTICELLA URCEOLARIS.

GENERIC CHARACTER.

A small animal, with a vascular cup; the mouth is at one end ciliated, and capable of being contracted; the stem fixed.

SPECIFIC CHARACTER.

Single, with a short tail, and toothed mouth.

This Animalcum is but perceptible to the naked eye, appearing as a small white speck; the microscope discovers the external covering to be so transparent, that all the motions of the animal within are perfectly distinct. It hath a double rotatory instrument, which, however, it can conceal or shew at pleasure; and it hath power to protrude the head and tail as at FIG. 1. or to contract them within the external coat or covering, as at FIG. 2.—When the animal intends to display its rotatory instrument, it forces its tail through the hole at the extremity of the outer coat, and affixes it to whatever substance is near; but when it swims, it moves its tail backwards and forwards to assist it.

They are found in river, or stagnant, water.

G

FIG.

PLATE XVII.

FIG. III.

TRICHODA VERMICULARIS.

GENERIC CHARACTER.

An invifible, pellucid, hairy worm.

SPECIFIC CHARACTER.

Long cylindrical trichoda, with a fhort neck, the apex hairy.

Is found in river water, and can affume various forms, as in our figure fhewn.

PLATE

18

PLATE XVIII.

NEPA CINEREA.

WATER SCORPION.

HEMIPTERA.

GENERIC CHARACTER,

Antennæ, or fore legs, cheliform, wings croſſed and complicated; fore part coriaceous.

SPECIFIC CHARACTER.

Black brown. Head ſmall. Antennæ cheliform. Thorax almoſt ſquare. Target large, brown. Shells large. One joint in each foot. Length near an inch. In the female the abdomen terminates in two long appendices. Four legs only.

There are three ſpecies only of this genus common to our waters, though the waters of hot countries abound with various kinds, ſome conſiderably exceeding in ſize even our *Sphinx Atropos.*

The Inſect ſinks its eggs into the cavity of a ruſh, or other aquatic plant, whence the larvæ are hatched. The perfect inſect is found in

4   June,

June, and thence to September or later; they are voracious, and feed on other aquatic animals, grasping their prey between their fore feet, and tearing them to pieces with their sharp rostrum.——They fly in the evening, and thus remove in herds from one pool to another when danger approaches.

It is supposed by some authors, that the fore feet of the nepa are the antennæ, and if this be admitted, the Insect hath only four feet; but if considered destitute of its antennæ, it hath six.

PLATE XIX.

CHRYSIS BIDENTATA.

HYMENOPTERA.

GENERIC CHARACTER.

Thorax joined to the abdomen by a fhort pedicle. Abdomen divided into three fegments. Sting fimple. Wings not folded. Antennæ filiform of one long and eleven fhort joints each.

SPECIFIC CHARACTER.

Head and laſt fegment of the body, fky blue, changeable. Thorax, and two firſt annulations of the abdomen, crimfon with gold fpots. Thorax with two teeth.

The Chryfis Bidentata is fcarcely fo large, and by no means fo common as the Chryfis Ignita, (not exceeding one-third of an inch in length) but is equal, if not fuperior in beauty and richnefs of colour. The head, but more particularly the laſt fegment of the body, appears in one direction of light, blue, in another green, in another purple, &c. and the thorax, and two firſt fegments of the abdomen are far more enriched with a golden appearance; the ground colour is deep crimfon, but the metallic appearance on the lighter parts, and the number of fmall gold fpots which befprinkle it, greatly diminifh the ſtrength of colour, and renders it, even before it is magnified, a fuperb little infect.

It is found in May or June in fome parts of Kent and Effex.

H   PLATE

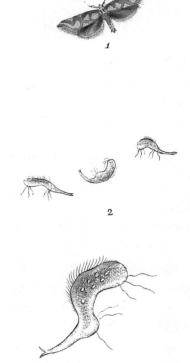

1

2

[45]

PLATE XX.

FIG. I.

PHALÆNA CHRISTIERNANA.

LEPIDOPTERA.

GENERIC CHARACTER.

Pyralis.

SPECIFIC CHARACTER.

First wings yellow, with rose-coloured marks. Under wings flossy, brownish grey.

————————————

The natural history of this Phalæna is so little known, that we freely confess our inability to shew its transformations; and although a deviation from our first intention, we trust the scarcity of the Fly will plead our excuse. We shall occasionally introduce figures of some rare and non-descript subjects, which we hope will be highly satisfactory to many of our subscribers.

Our specimen was taken at Feversham; they are sometimes met with about Darnwood in June or July.

FIG.

F I G.   II.

H I M A N T O P U S   L U D I O.

GENERIC CHARACTER.

A pellucid, invifible, cirrated worm.

SPECIFIC CHARACTER.

Curled Himantopus; the upper part hairy, the tail extended upwards.

PLATE

PLATE XXI.

PHALÆNA PRUNARIA.

ORANGE MOTH.

LEPIDOPTERA.

GENERIC CHARACTER.

Geometræ. Antennæ feathered.

SPECIFIC CHARACTER.

Wings orange, fprinkled with brown, and a femi-lunar fpot on the firft pair. Female paler than the male. Caterpillar yellow brown, with two fpines before and two behind.

―――――――――

The caterpillars of this Moth feed on fruit-trees, or on thorns, in the month of May; the Chryfalis is commonly found, rolled up in a decayed leaf, inwardly protected by the web, in June; and in July the Moth.

The prefent figure is of the male.

PLATE

PLATE XXII.

TIPULA PLUMOSA,

SEA TIPULA.

DIPTERA.

GENERIC CHARACTER.

Head long. Palpi four, curved. Trunk very fhort.

SPECIFIC CHARACTER.

Brown. Thorax greenifh. Eyes black. Fore legs longeft. Wings fhorter than the abdomen.

———————

Is found in the month of April near marfhes, and has been fre-quently miftaken for the common Gnat.

PLATE XXIII.

SILPHA VESPILLO.

COLEOPTERA.

GENERIC CHARACTER.

Antennæ clavated, foliated. Head prominent. Thorax margined.

SPECIFIC CHARACTER.

Margin of the thorax broad. Shells abbreviated, black, with two orange belts. Thigh of the hind legs large, with a spine near their origin; length one inch.

———————————

This species, like most of the Coleopterous Insects, delights in filth and putrefcence, and are rarely found except in the dung, or dead bodies of larger animals, whose entrails they devour; they prey on the larvæ of smaller insects beneath the surface of the earth, or they will destroy each other. Their Grubs are secreted in perforations made in the earth by the female, and therein they change to their last or perfect state in June or July: those Grubs are to be found by following the track of a plough.

They fly well with the transparent wings, which are concealed beneath the Elytra or upper Shells. The male is rather smaller than the female, and the orange belts are of a deeper hue: though both male and female vary in the strength of colour when alive, and yet more when preserved in cabinets, as they sometimes become almost brown. All insects are subject to this change, whatever may be the care of the collector.

I   PLATE

PLATE XXIV.

LIBELLULA DEPRESSA.

DRAGON FLY.

NEUROPTERA.

Wings four, naked, tranſparent, reticulated with veins or nerves. Tail without a ſting.

GENERIC CHARACTER.

Mouth with two long lateral jaws. Antennæ very ſhort. Tail of the male forked. Wings extended.

SPECIFIC CHARACTER.

Eyes brown. Thorax greeniſh, with two yellow tranſverſe bands. A large black ſpot at the baſe of each Wing, and a ſmall dark mark on their exterior margin. Body depreſſed, lance-ſhaped.

All the ſpecies of Libellula, but particularly the larger kinds, are conſidered by many rather as objects of terror, than ſubjects worthy inſpection; and the vulgar denomination of *Horſe-ſtinger*, contributes to this abhorrence: although it hath no power over animals of ſuch magnitude, it is perfectly a Vulture among lepidopterous, or other defenceleſs Inſects, deſtroying more for its ſport than for its voracious appetite.

The Fly is on the wing in May, and June, in almoſt every marſhy ſituation; the female lays her eggs near the roots of Oſiers on the banks of ditches, or ſinks them into the ſtalks of Ruſhes in the water; they hatch, and an ugly apterous inſect, of a brown colour, comes

forth;

PLATE XXIV.

forth; it hath a long body like the Fly, fix Legs, and a forked Head, a fharp fpine at the extremity of the abdomen, and a row of fpines on each fide, one at every joint; it plunges into the water, and immediately devours fuch of the inhabitants, or their eggs, as comes within its reach, and it continues this life of depredation until its next change. They are to be taken with a, fmall hand-net.

All tranfparent objects, in a certain direction before a microfcope, reflect the colours of the prifm. The *Tipula Plumofa* exhibits, in this fituation, an effulgence of colouring, which its natural fize conveys but fmall veftiges of; and the colours on the wing of this Libellula appears far more vivid when magnified.

The body of the male is bluifh grey; the prefent fpecimen is the female.

PLATE

PLATE XXV.

SPHINX API-FORMIS.

BEE HORNET SPHINX.

GENERIC CHARACTER.

Antennæ thickeft in the middle. Wings, when at reft, deflexed.

SPECIFIC CHARACTER.

Wings tranfparent, with brown veins. Abdomen yellow, the firft and fourth divifion from the thorax dark, purplifh. Thorax brown, with two yellow patches in front. Head yellow. Antennæ dark brown. *Linn. Syft. Nat.*

The Caterpillar of the *Sphinx Api-formis* is an internal feeder, and found only by making an incifion into the innermoft fubftance of the Poplar, the only tree the female commonly depofits her Eggs on; it is to us unknown, as is alfo the time of contmuing within the trunk of the tree; but in June, early in the morning, or in the evening, the Chryfalis is feen iffuing through the bark, from a perforation in the trunk, which the Caterpillar had formed previous to its change, gene-rally to the depth of fix or eight inches, or more. Nature has furnifhed every fegment of the Chryfalis with a double row of fharp teeth, or fpines, therewith it firmly attaches itfelf to the fides of the cavity, and, by repeated exertions to break from its prifon, gradually comes forth; thus, when it hath extricated itfelf from the tree, and the Chryfalis is fuppotted as in our Plate reprefented, the upper parts burft afunder with violence, and the infect rufhes forth to enjoy " the tem-perature of the fummer feafon." It is rarely found except in Effex.

K

There

There is another *Sphinx*, which differs in fo few particulars, that it hath been miftaken for the prefent fubject; notwithftanding, it may be eafily diftinguifhed by a crefcent of yellow in the fore part of the thorax, and thence entitled the *Lunar Hornet Sphinx*; a Drawing of which Infect, with the larva, has been prefented to the *Linnæan Society.* This larva is nearly the fize of the Buff-tip Caterpillar, and of an obfcure brown colour; probably the larva of the *Sp. Api-formis* may much refemble it.

It is arranged in many cabinets under the title of *Sphinx Vefpi-formis*; but the *Sp. Vefpiformis*, in the *Linnæan Collection*, now in the poffeffion of Dr. *Smith*, fcarcely exceeds half the fize of this fub-ject, and is probably unique. The *Lunar Hornet-Sphinx* had no place in that cabinet.

P L A T E

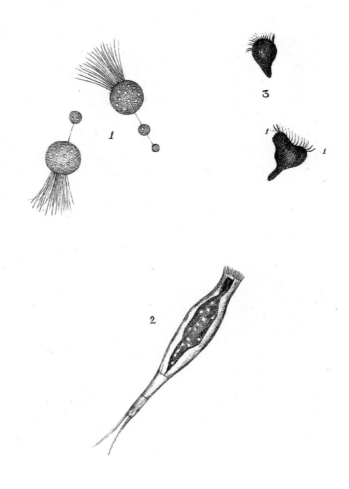

PLATE XXVI.

FIG. I.

TRICHODA COMETA.

GENERIC CHARACTER.

An invifible, pellucid, hairy Worm.

SPECIFIC CHARACTER.

Spherical, the fore part hairy, with an appendant globule.

FIG. II.

TRICHODA LONGICAUDA.

SPECIFIC CHARACTER.

Cylindrical, the firft part truncated, and fet with hairs. The tail long, with two joints, and terminated by two briftles.

FIG. III.

VORTICELLA TROCHIFORMIS NIGRA.

GENERIC CHARACTER.

A Worm, capable of contracting or extending itfelf, naked, with rotatory cilia.

SPECIFIC CHARACTER.

Top-fhaped black vorticella.

This

This fpecies of Vorticella appears, without the affiftance of a micro-fcope, as fmall black fpecks, fwimming on the water, particularly in meadows which are inundated. They are conftantly in motion; and two fmall white hooks are perceptible by glaffes at 1—1; by the help of thofe it is fuppofed to fwim, or they may inclofe fome rotatory or-gan. The infect is opaque.

PLATE

PLATE XXVII.

LEPTURA ARIETIS.

COMMON WASP BEETLE.

COLEOPTERA.

GENERIC CHARACTER.

Antennæ tapering to the end. Shells narrower at the apex. Thorax fomewhat cylindrical.

SPECIFIC CHARACTER.

Black. Anterior and pofterior margin of the Corflet yellow. Four yellow lines on each elytra or Shell. *Lin. Syft. Nat.*

They fly well, and are fometimes found on aquatic plants. They are exceedingly numerous in Kent, in the peafe and bean-fields, in May, or on the currant-bufhes, and not unfrequently are taken on the fern.

PLATE

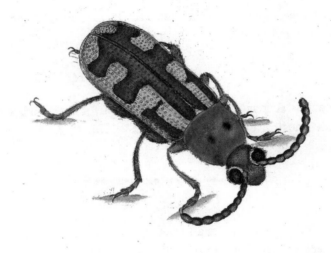

PLATE XXVIII.

CHRYSOMELA ASPARAGI.

COLEOPTERA.

GENERIC CHARACTER.

Antennæ knotted, enlarging towards the ends. Corslet margined, and body oblong. Thorax narrow.

SPECIFIC CHARACTER.

Head, Antennæ, and under side of the Body black. Thorax red, with two black spots. Shells dark green, with six yellow spots. Length one line. *Lin. Syst. Nat.*

This pretty Cloeopterous Insect is found in June on the Asparagus, when in seed. Linnæus calls it *Asparagi*, from the larvæ feeding on the leaves of that plant. It is a common insect, but forms a beautiful opaque object for the microscope. The natural size is given at Fig. I. and the magnified appearance above.

PLATE

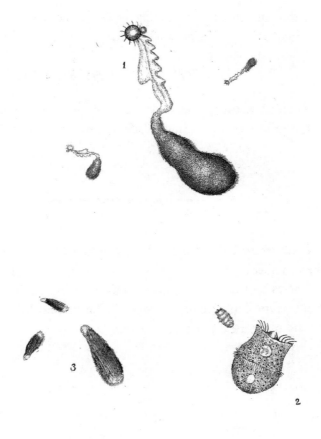

PLATE XXIX.

FIG. I.

TRICHODA MELITEA.

GENERIC CHARACTER.

An invifible, pellucid, hairy Worm.

SPECIFIC CHARACTER.

Oblong ciliated trichoda, with a dilatable neck, the apex globular, and furrounded with hairs. *Müller's Animalcula Infuforia,* &c.

Invifible to the naked eye, and rarely found except in falt-waters, although we have met with one fpecimen in the water of the Thames.

FIG. II.

VORTICELLA NASUTA.

GENERIC CHARACTER.

A Worm, capable of contracting or extending itfelf, naked, with rotatory cilia.

SPECIFIC CHARACTER.

Cylindrical, with a prominent point in the middle of the cup. *Müller's Anim. Infuf.*

Is invifible to the naked eye, and appears of an unequal fize before the microfcope is pellucid, with the fore part truncated and ciliated, and moves in the water with great alertnefs, by the affiftance of the circle of hairs which encompafs the body.

L   FIG.

F I G.　III.

V O R T I C E L L A　V I R I D I S.

GENERIC CHARACTER.

A worm capable of contracting or extending itfelf, naked, with rotatory cilia.

SPECIFIC CHARACTER.

Cylindrical uniform, green, and opake.　*Müller's Anim. Infuf.*

———————————————

The naked eye difcovers this fpecies as a mere point: when magnified it is of a dark green colour, almoft opake, nearly cylindrical, obtufe at the extremities, and deftitute of limbs. It moves circularly, or in a ftrait direction, and caufes fuch an agitation of the water, that notwithftanding its appearance, fome rotatory inftrument muft be concealed within the body, which the infect can put forth at pleafure.

P L A T E

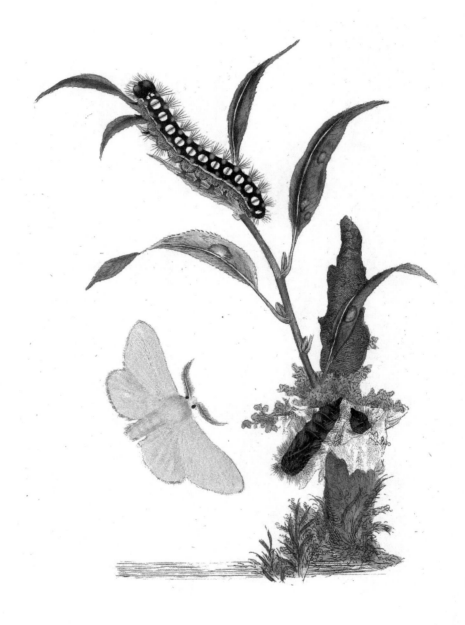

PLATE XXX.

PHALÆNA SALICIS.

WHITE SATTIN MOTH.

GENERIC CHARACTER.

** No trunk, wings depreſſed, deflexed, back ſmooth.

SPECIFIC CHARACTER.

Antennæ feathered. Body and wings white.
Caterpillar black, with red and white ſpots.

Are very numerous in the adjacent parts of London, and are found in the ſtate of Caterpillar, Chryſalis and Moth, at the ſame time, though commonly the Caterpillar changes to Chryſalis in June, and to a Fly in July.

It feeds on the Willow, the Ozier, the Poplar, &c.

31

PLATE XXXI.

FIG. I.

MUSCA CHAMÆLEON.

DIPTERA.

Two Wings.

GENERIC CHARACTER.

Musca, a soft flexible Trunk with lateral Lips at the end. No palpi.

SPECIFIC CHARACTER.

Dark brown or black. Antennæ taper, broken. Eyes large. Abdomen nearly circular, with three triangular yellow spots on each Side, and one at the extremity.

Linnæus, in a former edition of the *Fauna Suecica* gave this insect the name of *Oestrus Aquæ*, but he afterwards discovered it to be a Musca, and called it Musca Chamæleon. It is one of the most common Dipterous, or two-winged Insects we have; yet though so well known in its perfect state, few have attended so minutely to its changes as to discover that; they form the most singular part of its history.— The female deposits her eggs in the hollow stalks of aquatic plants, or broken reeds, or so provides for them that they cannot, but by some unforeseen accident, be carried away. The egg, in due time ripening, produces a Larva, no way resembling the Parent, but rather a Worm

M   of of a moſt ſingular ſtructure. This happens about the latter end of *May*, or beginning of *June*, if the weather proves favourable; they will then be found in ſhallow ſtanding waters, crawling on the graſs or plants which grow there, or they may be taken floating on the ſurface of the water. The Body conſiſts of twelve annular diviſions, whereof the Head and Tail are two; the Tail has a verge of hairs, which, when entirely expanded, ſupport the creature on the ſurface, with its head downwards. If it wiſhes to deſcend, it contracts the hairs in the form of a wine glaſs, or entirely cloſes them at the end; and when again it is riſing to the ſurface, it forces a bubble from a ſmall aperture in the center, which immediately makes a paſſage for its aſcenſion.— It changes to the Pupa ſtate, and about the middle of *July* to the Fly. It ſubſiſts at this time on the nectar and other juices it extracts from the bottom of the corolla in flowers.

F I G.

PLATE XXXI. 69

FIG. II.

MUSCA PENDULA.

DIPTERA.

MUSCA.

SPECIFIC CHARACTER.

Head black. Thorax yellow, with three longitudinal black lines. Abdomen yellow, with tranfverfe black marks.

Its habits nearly correfpond with thofe of the *Mufca Chamæleon.* Like that Infect it once wore the appearance of an Aquatic, and like it alfo in its laft or perfect ftate, exifts by extracting with its Trunk the nectar from flowers. It is to be taken in *June.*

FIG. III.

MUSCA LATERALIS.

DIPTERA.

MUSCA.

SPECIFIC CHARACTER.

* Thorax black. Abdomen bright red or brown, with a line of black from the Thorax; the laft fegment black, with hairs or fpines.

Vifits flower gardens in the month of *June.*

M 2   **PLATE**

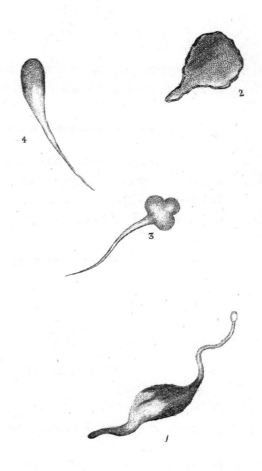

32

PLATE XXXII.

FIG. I.

VIBRIO OLOR.

GENERIC CHARACTER.

An invifible Worm, very fimple, round, and rather long.

SPECIFIC CHARACTER.

Elliptical, with a very long Neck, and a knob on the Apex.—— *Müller*'s *Ani. Inf.*

The Neck of this Creature is in continual motion, and the whole Body is dilatable. It is found in water, replete with decayed vegetables.

FIG. II.

KOLPODA MILEAGRIS.

GENERIC CHARACTER.

An invifible, very fimple, pellucid, flat, crooked Worm.

SPECIFIC CHARACTER.

Changeable, with the fore part like a hook, the hind part folded up.——*Müller*' *Ani. Inf.*

FIG.

FIG. III. and FIG. IV.
PROTEUS TENAX.

GENERIC CHARACTER.

An invisible, very simple, pellucid Worm, of a variable form.

SPECIFIC CHARACTER.

Running out into a fine point.—*Müller.*

A gelatinous pellucid body, stored with black molecules; it changes its form in a regular order, first extending itself out in a strait line, the lower part terminating in an acute bright point, without any inteftines, and the globules being all collected in the upper part, it next draws the pointed end up towards the middle of the body, swelling it into a round form. The contraction goes on for some time, after which the lower part is swelled as in Fig. IV. The point is afterwards projected from this ventricose part. It passes through five different forms before it arrives at that represented at Fig. IV. It scarcely moves from one spot, only bending about sideways. It is to be found in river water, where the *Nitida* grows.—*Adams on the Microscope.*

PLATE

33

1

2

PLATE XXXIII.

FIG. I.

PHALÆNA BATIS.

PEACH-BLOSSOM MOTH.

LEPIDOPTERA.

PHALÆNA.

GENERIC CHARACTER.

Antennæ taper from their apex. Wings in general contracted when at reft. Fly by night.

* NOCTUA.

SPECIFIC CHARACTER.

Firft pair of Wings brown, with five peach-coloured fpots on each. Second pair light brown.

———————————

The Peach-bloffom Moth at firft fight fo evidently diftinguifhes itfelf, that it can fcarcely be miftaken. The upper or firft pair of Wings have the ground of a brown colour, which in fome directions of light affume a golden appearance; and on each Wing are five elegantly difpofed fpots of white, having each a center of the moft beautiful bloom, or bloffom colour, which blend into the white with the moft exquifite foftnefs. The under Wings are of a fimple colour, and have only a tranfverfe fhade of a darker hue acrofs the middle of each Wing.

Its truly elegant appearance would alone be fufficient to claim our attention; but when we add that it is one of the rareft and moft

2

valuable

valuable fpecimens of Britifh entomology, it will be confidered as a compenfation for thofe more common fubjects occafionally introduced; and which the nature of our plan cannot permit us to refufe.

Our endeavours to procure the Caterpillar have hitherto been ineffectual, although it is very probably to be taken early in the feafon, feeding on the bramble. It is defcribed to be a brown larva, naked, or without hairs, with a gibbofity or rifing on the back, near the extremity.

Our Fly was taken in Effex, July 14th.

F I G. II.

P H A L Æ N A A M A T O R I O.

BLOOD VEIN, or BUFF ARGOS MOTH.

LEPIDOPTERA.

PHALÆNA.

** Antennæ feathered.

SPECIFIC CHARACTER.

Wings angulated, buff, fprinkled with brown, and a red tranfverfe line acrofs each. Margin of each Wing edged with red.

The Caterpillars of this Phalæna feed on the oak leaves. They are green, with yellow rings. The Fly is found in Effex very commonly in the month of July.

PLATE XXXIV.

FIG. I.

CURCULIO BACHUS.

COLEOPTERA.

Wings two, covered by two ſhells, divided by a longitudinal ſuture.

GENERIC CHARACTER.

Antennæ clavated, elbowed in the middle, and fixed in the Snout, which is prominent and horny. Joints four to each foot.

SPECIFIC CHARACTER.

Shells, and Thorax purple with gold ſhades; ſnout long, black. *Linn. Syſt. Nat.* 2. 611. 38. *Sckæff. Icon. Tab.* 37. *Fig.* 13. *Geoff. Inſ.* 1. 270. 4. *Sul. Hiſt. Inſ. Tab.* 4. *Fig.* 4.

Our figure repreſents the *Curculio Bachus*, as it appears before the Speculum of an Opake *Microſcope with a lens magnifying* times.

It is with this, as with many other ſpecies of inſects, and particularly thoſe of the Coleopterous Order, that unleſs they are in ſome meaſure magnified, much of their beauty will remain hidden, and much of their ſtructure be enveloped in obſcurity. It is not perfectly agreeable to our plan, and may admit of ſome blame from our ſubſcribers; but when objects ſo diminutive in ſize, and ſo complex in colour, offer to our attention, and it is not poſſible to repreſent them in their natural appearance, or in a manner ſatisfactory to ourſelves, we muſt have recourſe to the Microſcope for aſſiſtance. We conſider the confidence at preſent repoſed in our accuracy, and attention, to the natural ſubjects, evident from the general patronage beſtowed on our attempt; it is a ſpur to our exertions, and we will endeavour, as well by our future, as preſent correctneſs, to deſerve a continuation of

N

that

that efteem, and encouragement, fo liberally fhowered on our once arduous undertaking.

C. Bachus is near   in length, the Shells and Thorax appear of a deep gloffy purple, with much inclination to gold; a green and golden hue is feen on every part of the body as it moves in various directions of light. The whole appears before the microfcope befprinkled, and fpotted with gold and purple; gold in thofe parts where the light is moft powerful, and purple in the fhadows. The Snout is black, or of a dark colour, as are alfo the Eyes; and the fingular ftructure of the jointed Antennæ, which are thereon, deferve particular notice. This beautiful infect is as rare, as it is fuperb, and the larva is fcarcely, if at all known.—Our fpecimen was taken in the middle of *June*, in a field near Kent.

FIG. II.

CUCULIO GERMANUS.

BLACK CURCULIO.

COLEOPTERA.

Curculio.

SPECIFIC CHARACTER.

Snout long, black Head, Thorax, Shells and Body black. Two fmall fpots of yellowifh white on the fides of the Thorax.

Linn. Syft. Nat. 2. 613. 58.   *Scopol. Ann. Hift. Nat.* 5. 91. 44. *Frifch. Inf.* 13. 28. Tab. 26.

An Infect found in abundance in Germany, and by no means uncommon in this and every other part of Europe. It is generally taken in *June*.

I

PLATE

35

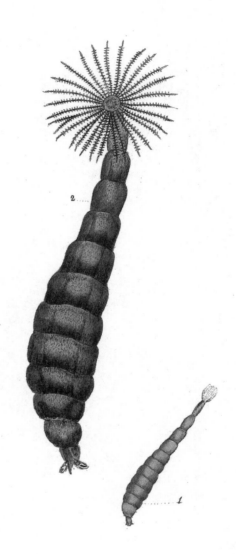

PLATE XXXV.

FIG. I.

Shews the natural fize of the larva, of the

Musca Chamæleon,

Defcribed in Plate XXXI of this work.

FIG. II.

As it appears magnified. We have taken it fince that plate was finifhed, or we would have introduced it with the Fly. Being unacquainted with any perfect reprefentation of this aquatic larva, we are happy to give it before the completion of the firft volume.

N 2   PLATE

36

P L A T E   XXXVI.

F I G.   I.

L I B E L L U L A   P U E L L A.

Neuroptera.

Wings four, naked, tranfparent, reticulated. With Veins or Nerves. Tail without a fting.

GENERIC CHARACTER.

Mouth with two long lateral Jaws. Antennæ very fhort, tail of the male forked. Wings extended.

SPECIFIC CHARACTER.

Body Saxon-greenifh blue, Eyes diftant, remote. Wings of equal length, with a cloud of brown in the middle, and without marginal fpot. Length two inches.

Linn. Syft. Nat. 2. 904. 20. *Fan. Sv.* 1470.

———————

It is neither fo large as to infufe terror by its appearance, or fo beautiful as to claim the firft place in a collection of Britifh infects: notwithftanding there are many inferior to it both in elegance, and colour. The whole of the body is a deep purplifh blue, which reflects on one part, a moft brilliant colour with a greenifh caft, and the clouds on the wings contribute much to its luftre. The Thorax and Head are nearly the fame, fome few fhades of green excepted.

It

It is found in *May* and *June*, sporting on the waters, or among the bushes which overgrow the sides of pools, or gently flowing streams; at noon, or after a shower, when the sun breaks from its watery prison, and penetrates the thickets, and the groves with inviting warmth, they are seen issuing from the dark retreat, and overhanging shrubbery; to bask and wanton in its effulgent beams, and fan the gently rising breeze with their lucid Wings. In many parts on the banks of the *Thames* they heighten the scene by the glow and richness of their colouring; the green, the blue, and the red; the yellow, purple, and the brown, in their richest teints, according to the species; and as they fly in various directions, display themselves in all their native elegance and splendor.

F I G. II.

L I B E L L U L A P U E L L A.

NEUROPTERA.

LIBELLULA.

SPECIFIC CHARACTER.

Body red with yellow and black lines at each segment; thorax green with yellow stripes. Wings clear, with marginal spots.

The body is red, with a yellow band and black mark at every segment; the Thorax green, with longitudinal lines of yellow; the Wings are perfectly transparent, except a marginal spot on each. It is voracious, as are all the species of Libellula, whether in the larva or the winged state; it appears about the same time as the preceding, and is the produce of an aquatic larva.

PLATE

I N D E X

V O L. I.

COLEOPTERA.

FIRST ORDER.

HEMIPTERA.

SECOND ORDER.

LEPIDOPTERA.

THIRD ORDER.

Phalæna

INDEX.

NEUROPTERA.

FOURTH ORDER.

HYMENOPTERA.

FIFTH ORDER.

DIPTERA.

SIXTH ORDER.

APTERA.

INDEX.

APTERA.

SEVENTH ORDER.

INDEX.

SPECIFIC NAMES,

ALPHABETICALLY ARRANGED,

TO

VOL. I.

Lucanus,

I N D E X.

I

ERRATA to VOL. I.

PLATE XXXIV. for *magnifying times*, read *magnifying four times*
for 6. *Bachus is near in length*, read C. *Bachus is near four Lines in length*
FIG. II. for *Cuculio*, read *Curculio*

THE

NATURAL HISTORY

OF

BRITISH INSECTS;

EXPLAINING THEM

IN THEIR SEVERAL STATES,

WITH THE PERIODS OF THEIR TRANSFORMATIONS,
THEIR FOOD, ŒCONOMY, &c.

TOGETHER WITH THE

HISTORY OF SUCH MINUTE INSECTS

AS REQUIRE INVESTIGATION BY THE MICROSCOPE.

THE WHOLE ILLUSTRATED BY

COLOURED FIGURES,

DESIGNED AND EXECUTED FROM LIVING SPECIMENS.

By E. DONOVAN.

VOL. II.

LONDON:

PRINTED FOR THE AUTHOR,

And for F. and C. Rivington, Nº 62, St. Paul's Church-Yard.

MDCCXCIII.

S

THE

NATURAL HISTORY

OF

BRITISH INSECTS.

PLATE XXXVII.

PAPILIO IRIS.

EMPEROR OF THE WOODS, OR PURPLE HIGH FLYER.

LEPIDOPTERA.

GENERIC CHARACTER.

Papilio.

Antennæ clavated. Wings when at reſt, erect. Fly by day.

SPECIFIC CHARACTER.

Wings indented; above, purple; darker round the Edges, with ſeven diſtinct white Spots on the firſt Wings; on the ſecond, an irregular broad white Stripe, and a red Eye. Beneath, black, brown, and white.

Linn. Syſt. Nat. p. 476. P. cx.

The Papilio Iris is eſteemed among the beautiful, and placed with the rare of the *Engliſh Lepidoptera.* The curſory reader may not perceive that ſuperiority, particularly as many of the minute Inſects infinitely excel in real beauty and richneſs of colouring; but the ſcien-

B 2

tific will be ever ready to give it the first place as a British Papilio, and to those a figure of the Caterpillar and Chrysalis will be an acceptable acquisition. It derives the title of Purple High Flyer, as it very rarely descends to the ground; except in some few instances, it has never been taken but in the most elevated situations, and even those instances have been after a strong wind, or heavy rain: The tops of the loftiest forest trees afford it an asylum, and in the Caterpillar and Chrysalis state, it is preserved from the wanton cruelty of man, by the almost inaccessible height of its habitation. They feed on the Sallow, *salix caprea*, and the Caterpillars are obtained by beating the branches of the tree with a pole twenty or thirty feet in length; it is then but a necessary precaution to cover the ground beneath with large sheets to a certain distance, or the insects which fall, will be lost among the herbage.

It is in Caterpillar about *May* and *June*; it passes to the Chrysalis state, and in *July* or *August* is a Papilio.

The great difficulty and trouble to rear the Caterpillars, when found; and greater difficulty to take the Fly, has stamped a valuable consideration on it, and particularly so when fine, and a high price is but esteemed an adequate compensation for it if in good preservation. The male is smaller, but more beautiful than the female; the upper side of the wings of the female not being enriched with that vivid change of purple which the male possesses in such an eminent degree; but the underside of the female is far richer in the various teints of colour than the male: they are both beautifully spotted, mottled, and waved with brown, black, white, and orange. The Chrysalis is of a very delicate texture, much resembling thin white paper, and is tinged in several parts with a very lively purple hue which it borrows from the wings of the enclosed insect, and bears the characteristic mark of a Papilio, by being suspended from the tail, with the head downward.

PLATE

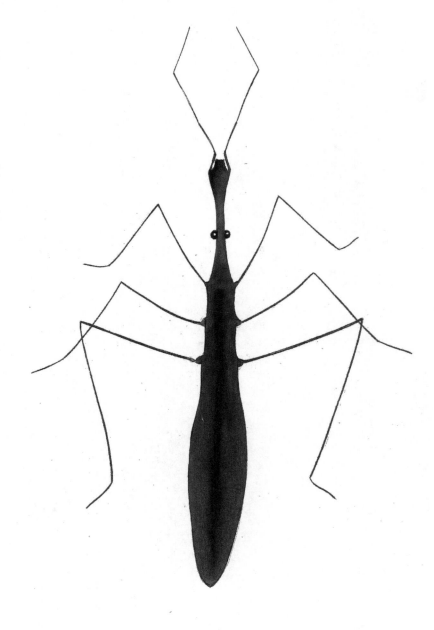

PLATE XXXVIII.

CIMEX STAGNORUM.

WATER BUG.

HEMIPTERA.

Shells or upper Wings femi-cruftaceous, not divided by a ftraight Suture, but incumbent on each other; Beak curved downward.

GENERIC CHARACTER.

Cimex Antennæ longer than the Thorax. Thorax margined. In each Foot three Joints.

SPECIFIC CHARACTER.

Black, brown, long, flender. Head one third of its whole length. Antennæ as long as the Head, and very flender. Eyes minute, prominent. Fore Legs fhorteft, length half an inch, breadth one third of a line.

Linn. Syft. Nat.

Many fpecies of the *Cimex* genus differ fo materially in their general form, that very nice attention is neceffary to difcriminate the fpecies which evidently belong to this extenfive family. The external appearance of the Houfe, or Scarlet Bug, cannot intimate the connection to the fame genus with this flender bodied infect; but fo they are arranged by Linnæus, and fo they will appear on a proper infpection of thofe parts which conftitute their generic character.

The prefent fpecies is common, and may be taken during great part of the warm feafons. We have an Infect of the fame genus (*Cimex Lacuftris*) which has frequently attracted notice by the variety and activity of its motions, when fporting on the furface of ftagnant pools, or other ftanding water : It appears to fly, or fkim the furface, but its wings are not often expanded, the lightnefs of its body

8   - 141 -   and

and length of its legs, permitting it to dart with great velocity in any direction, and when it alights, it caufes only a gentle tremulous motion beneath it. Its habits have much affinity to the generality of aquatic infects, and being conftantly found on that element, would almoft determine it to be of that race; but it is rather amphibious, and very rarely defcends beneath the furface. It will at intervals reft for feveral minutes motionlefs on the water, its fix legs are then expanded, and the tarfi of the feet only touch the furface; but the *Cimex Stagnorum* is remarkable for the regularity and carefulnefs of all its actions; it rarely runs, but treads the water, flow, and ever appears to apprehend danger; it frequently paufes fuddenly, and if it then perceives any thing difagreeable, retires. Aquatic Infects are generally fupplied at feveral parts of their body with an oily matter that the water cannot penetrate, and the legs of this Infect is apparently poffeffed of that property.

P L A T E

PLATE XXXIX.

COCCINELLA.

COLEOPTERA.

GENERIC CHARACTER.

Antennæ knotted, truncated. Palpi longer than the Antennæ; body hemifpheric. Shells and Thorax bordered. In each Foot three Joints.

FIG. I. and FIG. IV.

SPECIFIC CHARACTER.

COCCINELLA 22—PUNCTATA.

Head black, Corflet and Shells yellow. The firft with five black Spots, the latter with twenty-two. Length $1\frac{1}{2}$ line.

FIG. II.

14. PUNCTATA.

Shells orange, with fourteen black Spots. Head black. Thorax black in the Center, with an orange Margin and a black Spot on each Side.

FIG.

PLATE XXXIX.

FIG. III.

6. PUSTALATA.

Head, Thorax, and Shells black, with three red Spots on each Shell. Length 1½ line.

FIG V.

7. PUNCTATA.

Lady Cow, or Lady Bird.

Head and Thorax black, Shells red, with feven black Spots; length, three or four lines.

The hiſtory of thoſe ſeveral inſects ſo nearly reſemble each other, that one general account will compriſe all that can be ſaid of any of the ſpecies. The larva is not unlike the adult inſect, though its body is longer and tapering, and it hath no ſhells to defend it if in danger; its ſecurity therefore depends on its feet, which are rather longer, or at leaſt appear longer, than in the after-ſtate; all the ſpecies, whether as the larva or the adult, commonly feed on graſs, but they as frequently are taken on the plantain, thiſtle and roſe, or any other plant, whether wild or cultivated. They faſten themſelves to the leaves of any plant that is near when they enter the Chryſalis ſtate, and its appearance is then as if it were tied to the leaf by threads which paſs each other in tranſverſe directions; they remain only a few days in the Chryſalis, as it undergoes but little change. May, June, and July, or later if the weather ſhould prove fine, is the time to find them; many of the ſpecies are ſo numerous in almoſt every ſituation, that collectors give little trouble to obtain them, or at leaſt ſearch for ſuch only as are moſt uncommon.

PLATE

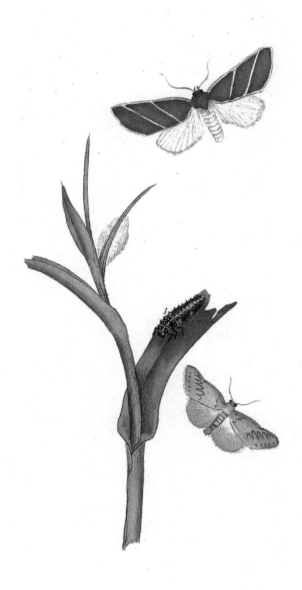

40

PLATE XL.

PHALÆNA ROSEA.

RED ARCHES.

LEPIDOPTERA.

GENERIC CHARACTER.

Antennæ taper from the base. Wings in general contracted when at rest. Fly by night.

SPECIFIC CHARACTER.

Rose colour. On the superior Wings a dark, waved, or arched line, and a row of spots near the margin.

———————

The Caterpillar of this *Phalæna* feeds on the Oak. Our specimen was taken from an oak at Norwood, July 15. They are not very common, although found, during the month of July, in several parts near London.

———————

FIG I.

THE LARVA

OF THE

COCCINELLA 7—PUNCTATA.

In Plate XXXIX we have represented several species of the *Cocci-nella* in their perfect or adult state. Our present figure is the larva of the 7 *Punctata*, Fig. V. It is a very common Insect; and will feed on almost every kind of vegetable food.

VOL. II. PLATE

PHALÆNA PRASINANA?

SCARCE SILVER LINE.

LEPIDOPTERA.

Phalæna.

SPECIFIC CHARACTER.

Body and under Wings white, firſt Wings green, with two oblique arrow lines of pale yellow.

––––––––––

We poſſeſs two ſpecies of the Green Silver Line; one Phalæna Praſinana, of Linnæus; the ſecond unknown to that author; but ſince deſcribed in the *Species Inſectorum* of *Fabricius.* Thoſe two ſpecies nearly reſemble each other, are both taken from the Oak, and are diſtinguiſhed only in ſome few particulars, the *Scarce Silver Line* has its Superior Wings of a plain pea-green, with two ſtripes of feint yellow, the Body and inferior Wings are of an immaculate white. But the Common Silver Line is more variegated in its colour, having a daſh of a paler hue between each Silver Line, and an orange or crimſon border. The Scarce Silver Line is taken in July, in woods.

Note, Fabricius appears to have changed the name of this Inſect in his *Spec Inſ.* for in the *Syſtem Entom.* he calls the common Silver Line Praſinana, the ſame as Linnæus does, which in the *Spec* he has altered to *Fagana.*

PLATE

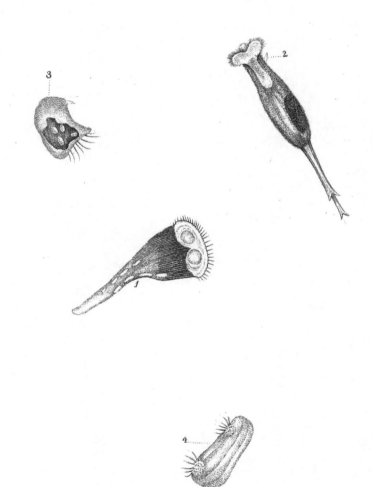

41

PLATE XL.

FIG. I.

VORTICELLA POLYMORPHA.

GENERIC CHARACTER.

A Worm, capable of contracting or extending itself, naked, with rotatory cilia.

Many-fhaped Vorticella green, opaque.

It is impoffible to defcribe the various forms thofe little Infects can affume; and, from the microfcope, it is both doubtful and difficult to give a correct figure of it, as the activity of its motions and changes frequently mifplace it from the verge of the focus. It is fcarcely perceptible to the naked eye, and is generally of a green colour.

FIG. II.

VORTICELLA ROTATORIA.

SPECIFIC CHARACTER.

Cylindrical Vorticella, with a little foot projecting from the neck; a long tail, furnifhed with four points.

Of all the fpecies of minute Infects, this Vorticella feems to have engaged the attention of the curious moft. Baker has defcribed

it

it under the title of *the Wheel Animalculum*, and hence it is well known. It is found in gutters, or leaden pipes, in the summer. This Insect possesses one property by no means common to larger animals, or even known of many of the minuter kinds; it lives in the water, but may be kept dry for months; and when again it is immerged in that element, it will regain its life and motion in half an hour.

F I G. III.

T R I C H O D A L Y N C E U S.

GENERIC CHARACTER.
An invisible, pellucid, hairy Worm.

SPECIFIC CHARACTER.
Nearly square; with a crooked beak. The mouth hairy.

F I G. IV.

K E R O N A P A T E L L A.

GENERIC CHARACTER.

An invisible Worm with horns.

With one valve, orbicular, chrystalline; the fore-part notched; the body lies in the middle of the shell: above and below are hairs or horns, of different lengths, jutting out beyond the shell, and acting instead of feet and oars.

Müller's Ani. Infus.

P L A T E

PLATE XLII.

FIG. I.

ICHNEUMON RAMIDULUS.

HYMENOPTERA.

Wings four; generally membraneous. Tail of the females armed with a fting.

GENERIC CHARACTER.

Ichneumon. Jaws, without tongue. Antennæ of more than 30 joints, long, filiform, vibrating. Sting within a bivalve fheath.

SPECIFIC CHARACTER.

Tawny brown. Thorax beneath, and extremity of the abdomen, black. Abdomen curved and compreffed.

FIG. II.

ICHNEUMON RAPTORIUS.

SPECIFIC CHARACTER.

Head, thorax, and extremity of the abdomen black; center fpot of yellow on the thorax; and two firft divifions of the abdomen bright roange. Legs black and brown.

4   Ichneumons

PLATE XLII.

Ichneumons are the moſt voracious of all the winged Inſects;—in their nature, robuſt and powerful, and armed with a formidable ſting; they are the dread, and deſtroyer of other tribes, and mortal enemies to each other; like the animal * whence their ſignificant appellation is derived, they exiſt by rapine and plunder, and ſupport their infant offspring on the vitals of larger Inſects.

The female Ichneumon, when ready to lay her eggs, is ſeen eagerly ruſhing from one plant to another, if its prey offers, which is generally the Larva of the *Phalæna, Papilio,* &c. it darts down with the ferocity of an eagle, and graſps the tender body in its claws; it is now in vain that the unwieldy animal attempts reſiſtance, as all its efforts are but the ſport of a ſavage conqueror. For raiſing the body almoſt upright, or into the form of a bow, the creature returns it in an inſtant, and daſhes the ſting up to the baſe, in the ſofteſt part of the caterpillar's body; this, if undiſturbed, it will repeat thirty or forty times, always chooſing a freſh ſpot for every new wound, and often entirely ſcarifying the Inſect. The ſtung animal refuſes to eat, and ſometimes its illneſs terminates in its death, though generally the eggs are matured, and the Inſects produced from the living body, ſo that if it ſurvives its miſery, and the wounds heal, the heat of the body ripens the embryos, and the young Ichneumons gnaw, and tear large paſſages through the body, to complete their delivery. July and Auguſt are the Months thoſe ſpecies we have deſcribed are on the wing.

* The Ichneumon is a well known animal in Egypt, particularly near the river Nile, and are uſeful for deſtroying the Eggs of the Crocodile, &c.

PLATE XLIII.

PAPILIO HYALE.

SAFFRON BUTTERFLY.

LEPIDOPTERA.

GENERIC CHARACTER.

Antennæ clavated. Wings, when at reft, erect. Diurnal.

SPECIFIC CHARACTER.

Wings entire, rounded, deep yellowifh orange. On the fuperior wings a black, and on the inferior wings an orange fpot in the center; and a deep irregular border of black on the margin. Antennæ and legs yellow. Breadth two inches.———*Syft. Ent.* 477. 148.—*Linn. Syft. Nat.* 2. 764. 100.—*Fn. Sv.* 1040.

———————

The *Papilio Hyale* has been defcribed by feveral authors, Englifh and Foreign, and the natural hiftorians of Germany have generally noticed it. Unlike many Infects we have in our country, it is found in every part of Europe, but in greater abundance in Africa and America.

VOL. II.   D   Its

Its breadth in England rarely exceeds two inches; but influenced by a warmer climate, they arrive at a higher degree of perfection than in thofe northern countries, at leaft they are commonly taken much larger. With us it has ever been efteemed as a rare Infect, though feen this feafon in Kent in greater plenty than for feveral years; but as they were probably only an accidental brood, they may again difappear for a confiderable time. The Fly is to be taken in autumn, but feldom after Auguft.

Our Figure is of the male;—the female has feveral irregular yellow fpots on the black borders.

PLATE

PLATE XLIII.

MELOE. PROSCARABEUS.

COLEOPTERA.

GENERIC CHARACTER.

Antennæ globular, the laſt globule oval. Thorax roundiſh. Shells ſoft. Head gibbous, and bent downwards.

SPECIFIC CHARACTER.

Blue, black. No wings. Shells ſhort. Abdomen long. Antennæ thickeſt in the middle. Head broad. Thorax narrower than the head, and without margin. Length $1\frac{1}{4}$ inch.————*Syſt. Ent.* 259. 1. ——*Linn. Syſt. Nat.* 2. 679. 1.—*Fn. Sv.* 826.

———————————

It is by no means for the beauty, but ſingularity of this creature that we have given it a place in our preſent ſelection. If it is too perfect for the larva of an Inſect, it certainly appears too imperfect for the adult ſtate; it has ſhells, but cannot fly, and their length compared with the proportion of the body contributes much to its awkward appearance. It is very quick-ſighted, and runs with ſwiftneſs when in danger. After death the body is conſiderably contracted, and the native brilliancy of colour it poſſeſſed while living immediately vaniſhes. When touched, a brown liquor oozes from the ſides.

We have ſeveral ſpecies of the Meloe differing in ſize, colour, and proportion; the *Meloe Proſcarabeus* is the moſt common, at leaſt near London. It feeds under the ſurface of the ground, on the tender fibrils of plants, and prefers the light earth of the flower-garden for its devaſtation. May be taken in May or June.

D 2   PLATE

44

PLATE XLIV.

THE LARVA

OF THE

LIBELLULA DEPRESSA.

In Plate 24 of this work we have reprefented the LIBELLULA DEPRESSA in the winged ftate, and our prefent Figure is, of the Larva of that Infect. We have before defcribed it as a favage vora-cious creature in every ftate of its exiftence. The Larva, which is an aquatic, feeds on Infects of that element; and when it becomes adult, Moths, Butterflies, and other winged Infects are its prey. As Lepidopterous Infects are not provided with any weapons, defenfive or offenfive, it will encounter the largeft, grafp them in its claws, and tear them to pieces. Its mouth is fpacious, and well adapted for that purpofe.

The Larvæ of moft winged Infects pafs to the Aurelia, or Chryfalis ftate, and thence produce the Fly; but the Larvæ of the *Libellulæ* never undergoes that change, and though its appearance is altered feveral times in its progrefs to perfection, it does not become dormant. When the ultimate period of its laft change arrives, it crawls to the bank, or fide of the ditch, and affixing its legs firmly to the ground, or grafs, it collects all its ftrength, and by one violent effort the fu-ture between the Thorax and Abdomen is broken, whence the Head and Thorax is protruded; after fome paufe the exuvia is caft off, and the Wings, which were before enwrapped in the fhort cafes at the bottom of the Thorax, expand. The creature now entirely formed for flight, only waits a fhort time to exhale the fuperfluous moifture, and then rufhes into the air, to fpread havoc and diforder.

PLATE

45

PLATE XLV.

PHALÆNA JACOBÆÆ.

CINNABAR MOTH.

LEPIDOPTERA.

GENERIC CHARACTER.

PHALÆNA.

Spiral Trunk; Back smooth, without Crest.

SPECIFIC CHARACTER.

Antennæ and body black. First Wings dark olive, with longitudinal red line near the anterior margin, and two red spots near the exterior. Second Wings red, with a black margin.——*Syst. Ent.* 588. 113.—*Linn. Syst. Nat.* 2. 839. 111.—*Fn. Sv.* 1155.

———————

As the Rag-wort grows spontaneously in almost every part of the country, the yearly increase of the Cinnabar Moth Caterpillars is generally considerable; and though many must inevitably perish before they arrive at perfection, the Fly may always be found in plenty in June, the Caterpillars in July and August.

PLATE

PLATE XLVI.

PHALÆNA FESTUCÆ.

GOLD SPOT MOTH.

LEPIDOPTERA.

GENERIC CHARACTER.

Spiral Trunk; Back fmooth, without Creft.

SPECIFIC CHARACTER.

Firft Wings brown, with two gold-filver fpots on each. Second Wings and Abdomen pale brown. Head. Antennæ and Thorax bright orange brown.———*Syft. Ent.* 607. 71.—*Linn. Syft. Nat.* 2. 845. 131.—*Fn. Sv.* 1170.—*Degeer Inf. Vers. Germ.* 2. 1. 312. 3.—*Albin. Inf. Tab.* 84. *Fig* G. H.—*Wilks Pap.* 8. *Tab.* 1. *a.* 17.—*Acta Holm.* 1748. *Tab.* 6. *Fig.* 3. 4.— *Kleman. Inf.* 1. *Tab.* 30. *Fig.* A.

The Caterpillars which are fmooth, and of a plain green colour, are found on fuch plants as grow in ditches, or fenny fituations.—The Sifymbrium Nafturtium, *Water Crefs,* is its common food, but it will devour with avidity moft aquatic vegetables, particularly the *Feftuca Fluitans, Floating Fefcue Grafs.* It is efteemed one of the rareft Species of Phalænæ we have in this country, its elegant form and rich colouring determines it alfo one of the moft beautiful. Near

VOL. II.   E   London

London it has been fought with moft fuccefs in the Batterfea Fields, or on thofe banks which abound with aquatic plants, between Batterfea and Richmond; the marfhes in the vicinity of Deptford and Rother-hithe have been yet more productive; we do not however underftand that any have been taken this feafon about the metropolis.

The very fingular manner in which this Caterpillar conftructs its web, deferves particular notice: previous to its transformation from the *Larva* to the *Aurelia*, it quits the tender plants which afford nou-rifhment, and retires to thofe, better calculated for its protection, in its defencelefs ftate; its choice is generally the Scirpus Lacuftris (*Bull Rufh*), or the ftouteft plant that is near, if its leaves are rufhy and ftrong. Its firft procefs is to make a deep incifion acrofs the leaf, which it effects with little labour, as its mouth is well armed for the purpofe; the upper part of the leaf being thus deprived of its fupport, inftantly becomes dependent; the Caterpillar embraces the two fur-faces of the fractured leaf, and weaves its web between. The web is of an exquifite texture and whitenefs, and bears great refemblance to the webs of fome fpiders that frequent watery places.

The Caterpillars are found in June and July, the Fly in Auguft.

PLATE

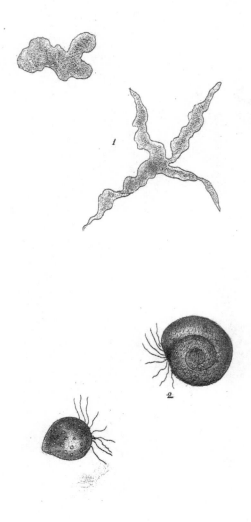

PLATE XLVII.

FIG. I.

PROTEUS DIFFLUENS.

GENERIC CHARACTER.

An invisible, very simple, pellucid Worm, of a variable form.

SPECIFIC CHARACTER.

Proteus, branching itself out in a variety of directions.

FIG. II.

TRICHODA BOMBA.

GENERIC CHARACTER.

An invisible, pellucid, hairy worm.

SPECIFIC CHARACTER.

Changeable, with a few hairs difperfed on the fore part.
Müller's Ani. Inf.

Proteus Diffluens, under fome of its changes appears rather a fhape-
lefs mafs, than an animated body; it confifts of gelatinous, pellucid
fubftance, replete with dark coloured molecules, which either direct or
attend, the internal exertions and actions of the animalculum; it pufhes
forth branches of various fhapes.

E 2

Is

Is found in fenny situations, but very rare; the author of the *Animacula Infusoria*, observed it only twice.

TRICHODA BOMBA.

Inconstant as the former, and nearly as difficult to define; it is sometimes spherical, immediately after it will become oval, Kidney shaped, &c. It is very lively, and darts with much velocity; is thick, pellucid, and of a clay colour, or brighter.

PLATE

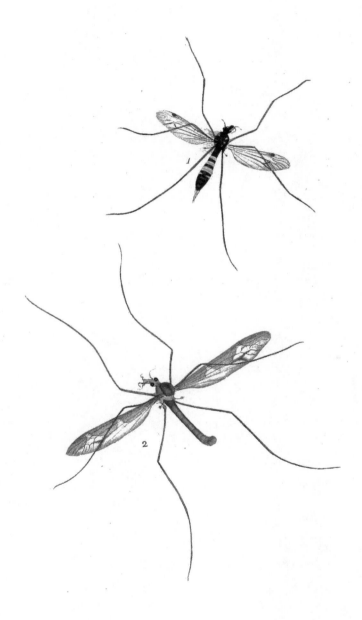

PLATE XLVIII.

FIG. I.

TIPULA CROCATA.

DIPTERA.

Wings **two.**

GENERIC CHARACTER.

Head long. Palpi 4, curved. Trunk very fhort.

SPECIFIC CHARACTER.

Black fpotted with yellow. Legs tawny, with black feet, and a black ring round the pofterior thighs. Wings tawny, with a marginal brown fpot.

> *Syft. Ent.* 748. 5.—*Linn. Syft. Nat.* 2. 971. 4.—
> *Fn. Sv.* 1739.
> *Geoff. Inf.* 2. 553. 7. *Tab.* 19. *Fig.* 1.
> *Degeer Inf.* 6. 349. 10.
> *Raj. Inf.* 72. 4.
> *Schaeff. Icon. Tab.* 126. *Fig.* 4.
> *Scop. carn.* 845.

FIG.

P L A T E XLVIII.

F I G. II.

T I P U L A R I V O S A.

SPECIFIC CHARACTER.

Brown-grey. Eyes black. Antennæ feathered. Wings larger than the body, with three brown patches near the margin. Tail of the female bifid. Length one inch.

Syft. Ent. 748. 2.—*Linn. Syft. Nat.* 2. 971. 2.—
    Fn. Sv. 1738.
Geoff. Inf. 2. 554. 2.
Degeer. Inf. 6. 341. 2. *Tab.* 19. *Fig.* 1.
Raj. Inf. 72. 2.
Scop. carn. 846.
Acta Holm. 1739. *Tab.* 9. *Fig.* 8.
Sulz. Inf. Tab. 20. *Fig.* 128.

The Genus *Tipula*, comprifes an extenfive family of the Dipterous Order, or of fuch Infects as are furnifhed with two Wings only. Our largeft Species are *Tip.* Rivofa, Crocata, Lunata, &c. the moft common is *T.* Oleracea, generally known by the trivial appellations, *Long Legs*, *Old Father*, &c. It is, as are alfo the other Species, perfectly harmlefs and inoffenfive; yet their fingular form, and more particularly the extraordinary difproportion of their legs, operates frequently to their difadvantage with the ignorant, who readily fuppofe they have to avoid, the fecreted fting, of whatever appears aukward or uncommon.

Our fmaller Species are infinitely more numerous, and many of them are not defcribed, being fo very minute as to remain unnoticed. The *Tipula Plumofa*, Plate XXII, differs materially in its general appearance from the larger kinds.

6

We

PLATE XLVIII.   31

We rarely find a specimen of the larger kinds of Tipula with the legs complete; the loss of one or two of those members do not materially retard the briskness of its motions, but it cannot fly after suffering a total amputation, though it will then live a considerable time.

The Tipula Rivosa being entangled by two of its legs in the snare of a large spider [ARANEA DIADEMA], at first endeavoured to disengage them by force, but this rather added to its calamity, and a third leg was attracted by the glutinous matter on the threads; the spider approached, and the creature accelerated its escape by leaving its legs in the web. It is very common to observe the broken limbs of the *Tipulæ* in the snares of this species of spider.

The Larvæ of many Tipulæ, more especially the very minute sorts, are found in standing water, but the larger, generally feed on the roots of grafs, and may be found by turning up the light surface of the earth. The Tipula Rivosa is taken in May and June, the Tipula Crocata in June and July; the latter is observed in the flower-garden or orchard.

PLATE

PLATE XLIX.

ARANEA DIADEMA.

WHITE CROSS, SPIDER.

APTERA.

No wings.

GENERIC CHARACTER.

Legs eight. Eyes eight.

SPECIFIC CHARACTER.

Abdomen gibbous, red-brown, with white spots in the form of a cross.

> Syst. Ent. 434. 13.—Linn. Syst. Nat. 2. 1030. 1.
> —Fn. Sv. 1993.

ARANEA cruciger.—Degeer. Inf. 7. 218. 1. Tab. 11. Fig. 3.

ARANEA Linnæi. —Scop. carn. 1077.

> Mouff. Inf. 233. Fig. 1.
> Aldrov. Inf. 608. Fig. 9.
> Tonst. Inf. Tab. 18. Fig. 17. 19. 20.
> Raj. Inf. 18. 2.
> List. Aran. Fig. 2.
> Frisch. Inf. 7. Tab. 4.
> Clerk. Aran. Tab. 21. Fig. 2.
> Schaeff. Elem. Tab. 21. Fig. 2.
> —— Icon. Tab. 19. Fig. 9.

The Genus *Aranea* includes a vast, if not endless variety of species, and though the greatest dissimilarity may be observed as to size, proportion, or colouring, of many individual kinds, yet the rapaciousness common to the family, is apparent in all. Our domeſtic Spiders are plain in their colours, and feldom attain a very extraordinary ſize; the gardens are infeſted by ſpecies ſomewhat larger, and more lively in their marks and teints, but if we wiſh to trace the juſt gradations of the beauty, or ſize, of thoſe deteſtable creatures, the foreſts abound; and will afford the higheſt gratification to the enquiries of the naturaliſt. We have Spiders purely white, or white ſtained with a lovely green; yellow, marked with a vivid red; purples ſhaded with the richeſt hues, and the brighteſt browns, beſpangled with the utmoſt elegance and ſymmetry : Yet under thoſe rich adornments which nature has ſo pro-fuſely beſtowed on this complication of beauty, and ferocity, we diſ-cover inherent qualities, which, in larger animals, would become formidable, and though we feel confident of our ſuperiority over the inſidious art of ſuch a contemptible creature, yet the mind is ſuſcepti-ble of an inward abhorrence at its touch, which neither the expanſion of philoſophy, or ignorance of its diſpoſition, will ſometimes ſuppreſs. It is probable, that *Thomſon*, in his deſcription of the Spider, felt this ſympathy of the human mind,

" ———— To heedleſs flies the window proves
A conſtant death ; where, gloomily retired,
The villain Spider lives, cunning and fierce,
Mixture abhorr'd ! Amid a mangled heap
Of carcaſes, in eager watch he ſits,
O'er-looking all his waving ſnares around.
Near the dire cell, the dreadleſs wanderer oft
Paſſes, as oft the ruffian ſhews his front ;
The prey at laſt enſnar'd, he dreadful darts
With rapid glide along the leaning line ;
And fixing in the wretch his cruel fangs,
Strikes backward grimly pleas'd : the flutt'ring wing,
And ſhriller ſound declare extreme diſtreſs,
And aſk the helping hoſpitable hand."

Early in the ſpring we find the neſts of Spiders in the crevices of old walls, trees, and other obſcure places. They are encloſed in webs of a white, yellow, or grey colour, varying according to the

ſpecies;

PLATE XLIX.

35

fpecies; immediately that the warmth of the fun has hatched them, they difperfe, it being no longer neceffary to live in focieties, which indeed, would deprive fome of their fubfiftence.

In February we took a neft of minute yellowifh eggs, which proved to be the infant offspring of the A. Diadema, they fcarcely exceeded the fize of a pin's head when hatched, and were of a bright yellow colour; at firft their food was the common houfe fly, but their increafe in bulk was fo rapid that it was neceffary to deftroy many, to preferve a few; we therefore felected four fpecimens, which being fed in feparate glaffes, and on different infects, exhibited each a diftinct degree of ftrength, and colour. One fpecimen deftroyed thirty of the common houfe fly in a day; it then appeared much enlarged, and the colours were almoft black, except the fpots of white, which fparkled with infinite luftre; but being confined a week without a frefh fupply, its colours were confiderably faded; another week of abftinence reduced its colours to a pale uniform brown, the body was much wafted, and the creature became perfectly ravenous. It devoured a vaft quantity of food, and recovered much of its former colours a few hours after.

Our largeft Spiders are incomparable for their fize, or venomous qualities, to the productions of America, or of the eaftern countries; in Germany they are far fuperior in fize to our fpecimens, but in Surinam they are infinitely furpaffed, Spiders of thofe parts being often found with legs as thick as a goofe-quill, and three or four inches in length, which with difficulty fupport a body as large as a pullet's egg. Their fnares are commonly extended from one branch of a tree to another, covering the fpace of twenty or thirty feet, and is fuffi-ciently ftrong to entangle the largeft infects. A. Seba has figured a Spider of this defcription, as defcending from an arm of a tree, into the neft of a fmall fpecies of Humming Bird, to fuck the blood of the parent, and eggs.

" The eyes of the Spider are a very beautiful microfcopic object, viewed either as tranfparent or opake; they have generally eight, two on the top of the head, that look directly upwards; two in the front, a little below the foregoing, to difcover what paffes before it; and on each fide a couple more, one whereof points fideways forward, the other fideways backward; fo that it can fee almoft all around it. They are immoveable, and feem to be formed of a hard, tranfparent, horny fub-ftance. The number of eyes is not the fame in all the fpecies of the

Spider.

Spider.   They have eight legs, with fix joints, thickly befet with hairs, and terminating in two crooked moveable claws, which have little teeth like a faw; at a fmall diftance from thefe claws, but placed higher up, is another, fomewhat like a cock's fpur, by the affiftance of which it adheres to it's webs; but the weapon wherewith it feizes and kills its prey is a pair of fharp crooked claws, or forceps, placed in the fore-part of the head.   They can open or extend thefe pincers as occafion may require; when undifturbed they fuffer them to lie one upon another.   Mr. *Lewenhoeck* fays, that each of thefe claws has a fmall aperture, or flit, through which he fuppofes a poifonous juice is injected into the wound it makes.

" The exuvia of the Spider, which may be found in cobwebs, being tranfparent, is an excellent object; and the fangs, or forceps, may be eafier feparated from it, and examined with more exactnefs than in a living Spider.   The contexture of the Spider's web, and their manner of weaving them, have been difcovered by the microfcope.   The Spider is fupplied with a large quantity of glutinous matter within it's body, and five dugs, or teats, for fpinning it into thread.   This fubftance, when examined accurately, will be found twifted into many coils, of an agate colour, and which, from its tenacity, may be eafily drawn out into threads.   The five teats are placed near the extremity of its tail; from thefe the aforefaid fubftance proceeds; it adheres to any thing it is preffed againft, and being drawn out, hardens in the air.   The Spider can contract or dilate at pleafure the orifices through which the threads are drawn.   The threads unite at a fmall diftance from the body, fo that thofe which appear to us fo fine and fingle, are notwithftanding compofed of five joined together, and thefe are many times doubled when the web is in formation."

——————— The Spider parallels defign,
Sure as Du Moivre, without rule or line.

POPE.

P L A T E

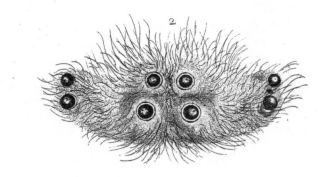

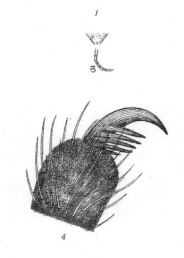

P L A T E L.

F I G. I.

The head and thorax, natural fize.

F I G. II.

A fragment of the head, with the eyes complete, as it appears when examined by the fpeculum of an opaque microfcope, defcribing the fituation of thofe organs, in this fpecies of Spider.

F I G. III.

One of its fore claws, natural fize.

F I G. IV.

The extremity of the claw magnified. Every foot is conftructed after this form.

PLATE

PLATE LI.

PHALÆNA PISI.

BROOM MOTH.

LEPIDOPTERA.

GENERIC CHARACTER.

Antennæ taper from the bafe. Wings, in general, contracted when at reft. Fly by night.

Noctua.

SPECIFIC CHARACTER.

Firft wings red brown, clouded with dark brown, two fpots in the centre, and a pale yellow undulated line near the exterior margin. Second wings and abdomen light brown with a broad fhade of a greyifh colour.

> *Syft. Ent.* 610. 88.—*Lin. Syft. Nat.* 2. 854. 172.—
> *Fn. Sv.* 1206.—*Degeer. Inf. Verf. Germ.*
> 2. 1. 322. 10.
> *Raj. Inf.* 160. 10.
> *Wilks pap.* 4. *Tab.* 1. *a.* 7.
> *Roef. Inf.* 1. *Phal.* 2. *Tab.* 52.
> *Merian. Europ. Tab.* 50.

The Caterpillars will devour indifcriminately the leaves of the knot-grafs, of peafe, the broom, &c. it is from the latter food, the Moth receives its name. The Caterpillars are found in July and

Auguft,

Auguſt, and deſcend into the ground late in September or the firſt week in October, and the Fly comes forth in July.

Caterpillars that enter the earth in the larva form, paſs to the chryſalis, and iſſue forth in the perfect or Fly ſtate, have no occaſion for a web to protect them ; and therefore few ſpecies prepare one. But among thoſe which remain expoſed in the open air, a very ſmall proportion neglect to weave a web with the utmoſt ſkill and induſtry ; the leaſt attentive to this apparently neceſſary precaution are the Papiliones, who, often regardleſs of their ſituations, are found [in chryſalis] ſuſpended againſt walls, the **trunks,** or branches of trees, and even paleings in very public roads.

PLATE

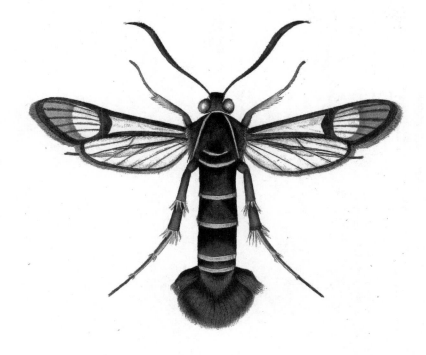

PLATE LII.

SPHINX TIPULIFORMIS.

CURRANT SPHINX.

LEPIDOPTERA.

GENERIC CHARACTER.

Antennæ thickeſt in the middle. Wings, when at reſt, deflexed. Fly ſlow, morning and evening only.

SPECIFIC CHARACTER.

Wings tranſparent, with black veins; a bright brown ſpot at the extreme angle of each ſuperior Wing. Abdomen, bearded; dark purpliſh black, with yellow bands.

> *Syſt. Ent.* 549. 9.— *Linn. Syſt. Nat.* 2. 804. 32.—
> *Fn. Sv.* 1096.
> *Clerk. Icon. Tab.* 9. *Fig.* 31.
> *Fueſl. Magaz. Tab.* 1. *Fig.* 6.
> *Harr. Inſ. angl. Tab.* 3. *Fig.* 8.

SESIA *Tipuliformis. Fab. Spec. Inſ. Tab.* 2. 157.

A very elegant, though common Species of the Sphinx *Genus:* it is taken in the months June and July. After the Inſect dies, the colour of the thorax and abdomen, except the yellow bands, is entirely black, or black with a very faint gloſs of a reddiſh blue: but is an exceedingly brilliant dark purple, while the creature is alive; and the yellow belts on the alternate diviſions of the body, glitter in the ſunſhine with the effulgence of molten gold. The legs are yet more beautiful, as the purple, though paler, is of a livelier luſtre; and every joint is deeply fringed with the ſame golden colour as that on the body.

The wings, which are perfectly tranfparent, except at the apex, are delicately veined, and ribbed with black lines. The fan tail is expanded or contracted at pleafure.

If the creature burfts from it's chryfalis in the morning, it is generally obferved fporting among the leaves of the neareft plants about noon; and this is commonly the time the male is feen feeking its mate.

It's very fingular appearance before the opaque microfcope, induced us to give the magnified figure, together with the Caterpillar, Chryfalis, and Sphinx, of the natural fize.

PLATE

PLATE LIII.

SPHINX TIPULIFORMIS.

CURRANT SPHINX.

CATERPILLAR, CHRYSALIS, and SPHINX of the Natural Size.

The Female depofits her eggs in the crevices of fuch twigs as are hollow; and a peculiar inftinct almoft invariably directs her to the ftalks of the currant trees: which are not only eafy of accefs, but afford grateful nourifhment to the young brood. Immediately that the Caterpillar is enlarged from the egg, it perforates the ftalk, and, having entire poffeffion of the inner channel, it feeds on the foft fubftance which is abundant within. Thus it is fecured by nature, with a defence againft many depredators, to which all Caterpillars, except internal feeders, are expofed.

It changes to a Chryfalis within the ftalk.

A fhort time before the Infect burfts forth, the Chryfalis is protruded through the outer bark, precifely in the fame manner as the Chryfalis of the *Sp. Apiformis* (PLATE 25.); and is fupported by a fimilar contrivance, every fegment being ferrated, or armed, with a row of very minute teeth, which firmly embrace the fubftance of the ftalk, and elevate the Chryfalis in an oblique pofture; until the laft efforts of the Infect completely difengages it from the cafe.

The *Sp. Tipuliformis* is the only Species of the tranfparent-winged Hawk-Moths, which is common near London; and is the fmalleft Infect of this divifion of the *genus:* the divifion contains few individual fpecies: but fuch as are generally very rare; at leaft the broods appear local in this country. The Currant Sphinx is taken in June.

PLATE

54

PLATE LIV.

CICADA.

HEMIPTERA.

Shells, or upper wings, femi-cruftaceous, not divided by a ftraight future, but incumbent on each other.   Beak curved downward.

GENERIC CHARACTER.

Antennæ taper.   Shells membraneous.   In each foot three joints. Hind legs ftrong for leaping.

FIG. I.

CICADA SANGUINOLENTA.

*   *   *

SPECIFIC CHARACTER.

Black.   Three red fpots on each fhell.

Syft. Ent. 688. 2.—*Linn. Syft. Nat.* 2. 708. 22.
Geoff. Inf. 1. 418. *Tab.* 8. *Fig.* 5.
Naturf. 6. *Tab.* 2.
Scop. carn. 330.
Fuefly. Inf. Helv. 24. 456.

CERCOPIS *Sanguinolenta.   Fab. Spec. Inf. t.* 2. 329.

The moft beautiful of the *Cicadæ* which inhabit this country; and rare with us, though common to many parts of Europe.   It is peculiar to the chalky and fandy foils of Dartford, and fome more diftant places.   It is taken in June and July.

FIG.

PLATE LIV.

FIG. II.

CICADA SPUMARIA.

CUCKOW-SPIT INSECT, or

FROTH WORM.

SPECIFIC CHARACTER.

Brown. Beneath lighter. Shells with two imperfect white belts, or long tranfverfe fpots, inferior wings pale.

Syft. Ent. 688. 5.—*Linn. Syft. Nat.* 2. 708. 24. *Fn. Sv.* 881.

Cicada fufca, fafcia duplici albida interrupta tranf-verfa.—*Geoff. Inf.* 1. 415. 2.

Cicada Spumaria Graminis fufca, alis fuperioribus maculis albis.—*Degeer Inf.* 3. 163. 1. *Tab.* 11. *Fig.* 1—21.

Locufta pulex Swammerdamii, nobis Cicadula. —*Raj. Inf.* 67.

Ranatra bicolor, capite nigricante.——*Petiv. Gazoph. Tab.* 61. *Fig.* 9.

Cicada fufca alis fuperioribus maculis albis, in fpuma quadam vivens.—*Degeer Acta Holm.* 1741. 221.

Vermes fpumans.—*Frifch. Inf.* 8. 26. *Tab.* 12.

Locufta germanica.—*Roef. Inf.* 2.—*Gryll. Tab.* 23.

Sulz. Inf. Tab. 10. *Fig.* 64.

Schaeff. Elem. Tab. 42.

Fuefly. Inf. Helv. 450.

CERCOPIS *fpumaria.*—*Fab. Spec. Inf. tom* 2. 329.

Cicadia

PLATE LIV. 47

Cicada Spumaria is not only common in this country, but is abundant in every part of Europe. It frequents moſt plants, but thoſe eſpecially which exhale much moiſture. The food of the Larva appears entirely of the vegetable kind, and conſiſts, for the moſt part, of the ſuperabundant fluids which all plants tranſpire.

" The CUCKOW-SPIT, or FROTH-WORM, is often found hid in that frothy matter which we find on the ſurface of plants. It has an oblong, obtuſe body; and a large head, with ſmall eyes. The external Wings, for it hath four, are of a duſky brown colour, marked with two white ſpots: the head is black. The ſpume in which it is found wallowing, is all of its own formation, and very much reſembles frothy ſpittle. It proceeds from the vent of the animal, and other parts of the body; and, if it be wiped away, a new quantity will be quickly ſeen ejected from the little animal's body. Within this ſpume, it is ſeen in time to acquire four tubercles on its back, wherein the wings are encloſed: theſe burſting, from a reptile it becomes a winged animal."

The colour of the winged Inſect is found to vary from a deep chocolate, to a very pale brown. It is taken in July and Auguſt.

It rarely uſes its wings for flight, as the hind legs are formed for leaping; at one effort it will frequently bound to the diſtance of two or three yards.

FIG.

FIG. III.

CICADA VIRIDIS.

*   *   *

SPECIFIC CHARACTER.

Head yellow, with two black fpots.   On the target two black dots.
Superior Wings green, with a yellowifh border.   Inferior Wings pale.
Body blue.   Legs yellowifh.

Syſt. Ent. 2. 685. 21.—*Linn. Syſt. Nat.* 2. 711. 46.
Fn. Sv. 896.

Locuſta pulex paullo minor.—*Raj. Inſ.* 68. 3.

Ranatra viridefcens. — *Petiv. Gazoph.* 73. *Tab.* 76. *Fig.* 6. —
Geoff. Inſ. 1. 417. 5.
Fueſly. Inſ. Helv. 24. 465.

CICADA *Viridis.*—*Fab. Spec. Inſ. t.* 2. 326.

———————

A fpecies not uncommon, but lefs plentiful than the *C. Spumaria.*
It is found in July and Auguft, on aquatic plants; generally on the
high rufhes which abound in marfhy places.

PLATE

PLATE LV.

PAPILIO URTICÆ.

SMALL TORTOISE-SHELL BUTTERFLY.

LEPIDOPTERA.

GENERIC CHARACTER.

Antennæ clavated. Wings, when at reſt, erect. Fly by day.

SPECIFIC CHARACTER.

Deep orange. Wings conſiderably indented. Above, on the ſuperior Wings, ſix black and two whitiſh ſpots. Inferior Wings, one large
ſpot on each. A broad exterior black border, edged with black and
yellow, and a row of light blue ſpots on each Wing. Underſide,
black-brown with waves and daſhes of yellow, brown, &c.

Syſt. Ent. 505. 263.—*Linn. Syſt. Nat.* 2. 777. 167.
—*Fn. Sv.* 1058.—*Geoff. Inſ.* 2. 37. 4.

Papilio urticaria vulgatiſſima, rufo nigro cœruleo et albo coloribus
varia. *Raj. Inſ.* 117. 1.

Robert. Icon. Tab. 5.
Huſſn. Pict. 2. *Fig.* 16.
Merian. Europ. 44. *Tab.* 44.
Albin. Inſ. Tab. 4. *Fig.* 51.
Schaeff. Icon. Tab. 142. *Fig.* 1. 2.
Goed. Inſ. 3. *Tab.* 3.
——————————— *&c. &c.*

VOL. II. H A very

PLATE LV.

A very beautiful fpecies of the *Papilio*; and, were it lefs frequent, would be infinitely efteemed for the elegant combination of its colours; but is at prefent little regarded. The old Flies are obferved in May, the Caterpillars are hatched about the middle of June; in July they are full fed, and caft their laft exuviæ : they transform into Chryfalis, in which ftate they remain only fifteen days, and then burft forth a Papilio.

They continue to breed in vaft quantities during the warm weather; and have, if the feafon be favourable, feveral broods before the winter.

The Chryfalis is brown, but often affumes much of a golden hue; and, though not its common appearance, is fometimes feen entirely of a rich gilded, or gold colour; but this is unnatural, and generally indicates that the Caterpillar has been ftung by the Ichneumon Fly. 'I he Caterpillars are taken on the Nettle *.

* *Urtica Dioica. Linn.*

PLATE

- 206 -

PLATE LVI.

FIG. I.

PYROCHROA COCCINEA *.

SPECIFIC CHARACTER.

Beneath, Legs and Antennæ black. Head, Thorax and Shells bright red, inclining to brown.

The above Infect which *Fabricius* has, after that celebrated French Naturalift Geoffroy, made a new genus, under the title *Pyrochroa* *, has in general been confidered by the Collectors of Infects as the *Cantharis Sanguinolentæ* of Linnæus; but this cannot be the cafe, as the defcriptions by no means correfponds; nor is it the *Lampyris Coccinea* of that author, as quoted by Fabricius; we are therefore inclined to think, that notwithftanding it is fo plenty with us, it was unknown to the Swedifh Naturalift at the time he wrote; efpecially a⸳ ᵗʰᵉ ᶠᵖᵉᶜⁱᵐ⸳ was not contained in his cabinet.

It is very common in England, in Jnly.

* *Geoff. Inf.* 1. 388. 1. *tab.* 6. *fig.* 4.

PLATE LVI.

FIG. II.

SILPHA QUADRIPUNCTATA.

COLEOPTERA.

GENERIC CHARACTER.

Antennæ clavated, foliated.   Head prominent.   Thorax margined.

SPECIFIC CHARACTER.

Head, Antennæ, and Legs black.   Thorax yellow, with a large
spot of black.   Shells yellow, with four small black spots.   Length
half an inch.

———————

Appears local to certain parts of this kingdom: is sometimes taken
by beating the Oaks in Caen-wood, near Hampstead, in July; it is,
however, rare.

1

2

3

PLATE LVII.

FIG. I.

PHALÆNA LAMDELLA.

TINEA.

SPECIFIC CHARACTER.

Superior Wings bright yellow brown, with a triangular dark fpot, extended obliquely from the inferior margin, to the center of the Wing, and terminated by a minute detached fpot of the fame colour.

A non-defcript, and has hitherto only been taken on Epping-foreft: the brood was difcovered in a furze-buih, by Mr. Bentley, an eminent Collector of Infects, in July 1789; the Cabinets of feveral Naturalifts have been fupplied from the parcel then taken, as the Species has rarely been obferved fince.

We prefer the name *Lamdella*, as the form of the Greek *Lamda* (λ) is well characterized, on the fuperior Wings.

FIG. II.

PHALÆNA AURANA.

PYRALIS*.

SPECIFIC CHARACTER.

Superior Wings brown, with two orange fpots on each; inferior Wings brown.

Syft. Ent. 653. 43.—*Fabri. Spec. Inf.* 11. 286. 66.

* *Fab. Gen. Inf.*

An

PLATE LVII.

An elegant Species of the minuter kinds of Lepidopterous Infects: it derives its name from the fpots of bright orange, or gold colour, which are on the fuperior wings: is very rare: our fpecimen was taken in Kent, late in July; it appears peculiar to that county only, or is certainly very unfrequently, if ever, found elfewhere.

Larva unknown.

FIG. III.

PHALÆNA APICELLA.

TINEA.

SPECIFIC CHARACTER.

Grey. A circular fpot of gold, or orange colour, at the apex of each fuperior Wing.

Non-defcript, and is alfo very rare. Our fpecimen was procured by beating a White thorn-bufh, on Epping-foreft, early in May.

The orange fpot on the ends of the upper wings afford the moft ftriking diftinction for a Specific Character; we therefore denominate it Apicella.

WISHING

WISHING to comprise such information as may recommend our Work, to a general Class of Readers, we are absolutely compelled to deviate from that uniform path which we at first intended to pursue; by introducing the figures of some Moths before we can procure their larva; we promise this will rarely occur, except with Insects whose larva are unknown; and the Author will spare no expence, or trouble, to attain even those: but, were he to refuse a place to the many valuable specimens recently discovered, it would be very displeasing to the greater part of his Subscribers; therefore, as an invariable observance of such intention, promises only to exclude the most rare of our Insects, we cannot always indulge it: on this plan, in the first Volume we could neither have represented the Phal. Batis, *Peach Blossom*, as the larva has only once been found; or the Phal. Christernana, whose larva is unknown*: These are Insects which few Cabinets in England possess; hence the figures must be very acceptable, and their rarity a sufficient apology for their premature introduction.

* The Caterpillars of a very small portion of minute Moths are known; and many Species in the adult state are so very rare, as to have escaped the attention of the most accurate Entymologists. Of the number which are ascertained as natives, very few are hitherto figured, or even described.

PLATE

PLATE LVIII.

FIG. I. I.

PHALÆNA PRUNIELLA.

LEPIDOPTERA.

GENERIC CHARACTER.

PHALÆNA.

Antennæ taper from the bafe. Wings, in general, contracted when at reft. Fly by night.

TINEA.

SPECIFIC CHARACTER.

Superior wings brown, inclining to purple; from the interior margin is extended a broad white dafh along the pofterior margin, nearly two thirds of its length; but is interrupted near the extremity by a fquare fpot of dark brown. Inferior wings grey. Head and thorax white. Abdomen grey.

We have copied the name *Pruniella*, from that celebrated work of Clerk, faid to be executed under the immediate infpection of *Linnæus* himfelf: He has figured it in the 11th Plate, Fig. 4. But the great fcarcity of that work, there not being twelve copies in this country, can have contributed in a very fmall meafure to its being generally known; which indeed is the fact, as it does not appear any writer fince that time has figured, or even defcribed it. Some were, perhaps, ignorant of its having been figured in Clerk's Plates, which however, could not have been the cafe with *Linnæus*; but we cannot find that he has defcribed it, or referred to Clerk's figure in any part of his works;

VOL. II.   I   though

though a copy of that book came over with the Linnæan collection, into the hands of Dr. Smith; nor can we trace any defcription of this moth in the writings of *Fabricius*; he alfo has not quoted the figure: We may hence conclude that although the infeɗ is frequent in the months of June and July, it is little known, except with thofe who poffefs colleɗions; and even many of that defcription are perhaps unacquainted with the circumftance of its having been named by Clerk, and probably by no other author.

Taken at Highgate.

F I G. II.

P H A L Æ N A M A R G I N E L L A.

L E P I D O P T E R A.

G E N E R I C C H A R A C T E R.

P H A L Æ N A.

T I N E A.

S P E C I F I C C H A R A C T E R.

Firft wings bright, pale brown, with a broad white margin. Second wings white.

Our prefent fpecies was unknown to *Linnæus*; but according to his definition of *genera*, is one of the *tineæ*; it will be neceffary, however, to diftinguifh it from the *tinea marginella* of Fabricius, which is a native of Germany, and altogether different; that writer, it is well known, divided many of the genera of Linnæus, and from their materials conftituted an infinitely greater number; it was by fuch divifions he feparated the *tineæ*, into the genera, *tineæ* and *alu-*

citæ, removing the *alucitæ* * of Linnæus under the title of *Ptero-phorus*.

He therefore ufes the fpecific name *marginella* to his *tinea* and *alu-cita*. Our fpecimen is defcribed by him, under the name *Alucita marginella*. It is found on the juniper in May.

Taken at Dartford.

F I G. III.

PHALÆNA PAVONANA.

LEPIDOPTERA.

GENERIC CHARACTER.

PHALÆNA.

TORTRIX.

SPECIFIC CHARACTER.

Superior wings clouded with black and buff-coloured markings, and a very minute reprefentation of a peacock's feather at the apex. A dorfal fpot of bright brown, furrounded with a deep black mark, Inferior wings grey brown, with the eye of the peacock's feather at the apex.

This fingular Tortrix, which abounds with beautiful markings, is particularly diftinguifhed by the elegantly little mark at the apex of the upper wings, which appears like the feather of a peacock's tail:

---

* The infects diftinguifhed by this title are known by the trivial names *Plumes*, or *Fans*; their wings being entirely formed of feathers connected only near the bafe in the manner of a fan.

the

the ferruginous dorfal fpot, furrounded with a thick black mark, although pretty, is by no means peculiar to this fpecies, being common to feveral other minute moths: the clouded markings of black and buff-colour, interfperfed with filver, give this little animal a beautiful appearance, particularly under the microfcope. The under wings have a fimilar appearance of a peacock's feather, but more obfolete at the apex.

We believe this fpecies has never been defcribed before, and very rarely taken. Our fpecimen was found in Suffex.—Auguft.

PLATE

PLATE LIX.

FIG. I.

PHALÆNA PAVONANA

MAGNIFIED.

FIG. II.

PHALÆNA PRUNIELLA

MAGNIFIED.

We cannot felect more pleafing objects for microfcopical inveftigation, than thofe two minute moths, efpecially the firft; the markings
appear rather confufed without the affiftance of glaffes, but a lens of
a very fmall power completely developes it of this imaginary obfcu-
rity, and difplays an elegance fufficient to recommend it to our attention; but independent of fuch confideration, it will, it is prefumed, be
confidered as a material advantage to the defcription annexed, to accompany the figure of the natural fize with a microfcopical reprefentation; not to enforce that fuch addition is indifpenfibly neceffary, but
when moths like the prefent offer, whofe marks, though beautiful,
appear confufed, it will certainly much affift to its neceffary information; as well as in future to determine the fpecies itfelf.

PLATE

PLATE LX.

CURCULIO SCROPHULARIÆ.

COLEOPTERA.

GENERIC CHARACTER.

Antennæ elbowed in the middle, and fixed in the fnout, which is prominent and hairy. Joints in each foot four.

*** *Long fnout. Thighs dentated.*

SPECIFIC CHARACTER.

Somewhat fpherical. Thorax narrow, befet with yellow-white hairs. Shells black brown, ftriated; a large black fpot on the future, on each fide of which are two fmall fpots. Length three lines.

Syft. Ent. 140. 68.—*Linn. Syft. Nat.* 2. 614. 61.
—*Fn. Sv.* 603.

Geoff. Inf. 1. 296. 44.
Degeer Inf. 5. 208. 3. *Tab.* 6. *Fig.* 17. 18. 19. 20?
Lift. Scarab. Angl. 395. 35.
Reaum. Inf. 3. *Tab.* 2. *Fig.* 12.

This fingular little infect feeds, when in the larva ftate, on plants of the *fcrophularia* genus, (fig-wort), and thence receives its fpecific name. The beetle is not uncommon in June, and is ufually found on the fame plants as the larva: the minutenefs of this creature evades a complete difcovery of the uncommonly teffelated appearance it affumes before the fpeculum of an opake microfcope; our plate reprefents the chryfalis and beetle, natural fize, together with a confiderably magnified figure of the latter.

PLATE

PLATE LXI.

PHALÆNA STRAMINEA.

LEPIDOPTERA.

GENERIC CHARACTER.

Antennæ taper from the bafe. Wings, in general, contracted when at reft. Fly by night.

NOCTUA.

SPECIFIC CHARACTER.

Antennæ and tongue deep yellow. Head and thorax covered with long hairs; which, with the fuperior wings, are pale yellow, or bright clay colour; in the middle of the fuperior wing is a kidney-fhaped fpot of dull grey, enclofed by a dark reddifh brown line, which is united to the anterior margin by another fpot of the fame colour. Near t e exterior margin is a broad obfolete band of pale brown, but where it touches the anterior margin it is darker; within this band are nine white fpots, or points, and between the band and exterior margin of the wing, on the lower edge, is a bright black point; there are feveral other reddifh brown points fcattered upon the upper wing, near the bafe. The inferior wings are of a yellowifh-white, with a fhade of purple, a dark fpot on the middle, and a pale black, broad border, with a white fringe.

This elegant fpecies of the *Noctua* divifion of Moths, appears to be not only a nondefcript, but altogether unknown before; even to the beft practical entymologifts: That an infect of fuch magnitude fhould have been unnoticed by *Linnæus*, or *Fabricius*, is not very fingular, as feveral nondefcripts of a fimilar, and many of an inferior,

VOL. II. K fize

fize, are to be feen in almoft every cabinet; but that the fpecies fhould have efcaped the refearches of the moft eminent collectors, is rather aftonifhing.

We have fought every information which our connection would permit; and from the refult we fcarcely hefitate to pronounce the infect of a nondefcript fpecies, and our fpecimen to be perfectly unique; at leaft it is a newly-difcovered acquifition to many fcientific entymologifts.

The original, whence the figure has been copied, is in the collection of the author; it was taken in a lane leading immediately from the wood at *Tottenham*, the laft week in June, 1793. It was difcovered in the evening, on a blade of grafs; and, from its wet appearance, as well as exquifite prefervation, it had certainly juft emerged from its chryfalis.

The Caterpillar may be fuppofed to be an underground feeder, and to fubfift on the roots of grafs, &c. or one of that kind which comes only above the furface of the earth in the night.

P L A T E

PLATE LXII.

MUSCA ONOPORDINIS?

DIPTERA.

Wings two.

GENERIC CHARACTER.

A soft flexible trunk, with lateral lips at the end.   No palpi.

SPECIFIC CHARACTER.

Head, thorax, and body, yellow brown.   Wings, variegated with brown spots.

Syst. Ent. 787. 80.
Fabric. Spec. Inf. 2. 455. 105.

Whether this is the Musca *Onopordinis* of Linnæus, as quoted, we cannot exactly determine; it answers to his description of that insect, but he speaks so very concisely, that we will not venture to assure ourselves of his M. *Onopordinis* being our species.   In this and many other instances we find, that though brevity is the greatest excellence of the Linnæan descriptions, it is also their most essential fault.

The species may, with much propriety, stand under the name *Onopordinis*, as we believe it has never been figured before; and, should the Linnæan species be hereafter discovered to differ from the present, a new name may be readily given to that insect.

Flies in April and May, and is very common in the summer, in woods.

PLATE

PLATE LXIII.

FIG. I.

SILPHA THORACICA.

COLEOPTERA.

GENERIC CHARACTER.

Antennæ clavated, foliated. Head prominent. Thorax margined.

SPECIFIC CHARACTER.

Black. Three longitudinal lines on each fhell. Thorax red-brown.

> *Syft. Ent.* 73. 6.—*Linn. Syft. Nat.* 2. 571. 13.—
> *Fn. Sv.* 452.—*Stroem. Act. Nidrof.* 3. *Tab.* 6.
> *Fig.* 1.
> *Silpha. Degeer Inf.* 4. 174. 3. *Tab.* 6. *Fig.* 7.
> *Peltis* nigra, &c.—*Geoff. Inf.* 1. 121. 6.
> *Scarabæus.*—*Raj. Inf.* 90. 10.
> *Cafida* nigra, &c.—*Gadd. Satag.* 25.
> *Silpha Thoracea. Scop. carn.* 54.
> *Bergftr. Nomencl.* 1. 23. 5. *Tab.* 3. *Fig.* 5.
> *Schaeff. Icon. Tab.* 75. *Fig.* 4.
> *Sulz. Inf. Tab.* 2. *Fig.* 12.

Taken at Charlton in June. It is a very rare fpecies in every part of this country, though not unfrequent in Germany.

FIG.

PLATE LXIII.

FIG. II. III.

CASSIDA CRUENTATA.

COLEOPTERA.

GENERIC CHARACTER.

Antennæ knotted, enlarging towards the ends. Shells and thorax bordered. Head concealed under the corselet.

SPECIFIC CHARACTER.

Bright green above, on each shell near the scutellum a very bright sanguineous mark. Beneath, body and thighs black. Legs and feet light brown.

Is found on verticillated plants and thistles in May.

Although confounded by some with the common Cassida (*C. Viridis*), it differs very essentially from that insect: it is smaller; of a deeper green colour, and does not fade to a dirty brown after death: but the bright sanguineous marks on the shells are scarcely visible in a dead specimen; the former is very common in May, but our species is rare.

C. Cruentata has never been either described or figured before.

FIG.

PLATE LXIII.

71

FIG. IV.

SILPHA OBSCURA.

COLEOPTERA.

SILPHA.

SPECIFIC CHARACTER.

Entirely black. Shells punctured; with three longitudinal lines on each.

Syst. Ent. 74. 11.—*Linn. Syst. Nat.* 2. 572. 18.
—*Fn. Sv.* 457.—*Scop. carn.* 57.
CASSIDA. *Udm. Diff.* 8.

Very frequent in May: breeds in corn-fields and meadows; but is found in many other situations.

PLATE

PLATE LXIV.

FIG. I.

CERAMBYX VIOLACEUS.

COLEOPTERA.

GENERIC CHARACTER.

Antennæ articulated, and tapering to the end. Shells long and narrow. Four joints in each foot. Thorax with lateral spines, or tubercles.

SPECIFIC CHARACTER.

Head, thorax, and shells, blue-purple. Legs, and underside black.

Linn. Syst. Nat. 2. 635. 70.—*Fn. Sv.* 667.

Degeer Inf. 5. 88. 24.

Stenocorus violaceus. *Scopol. Ann. Hist. Nat.* 597. 59.

Cantharis, &c. *Gadd. Diss.* 28.

Frisch. Inf. 12. *Tab.* 3.

Callidium violaceum. *Fab. Spec. Inf.* 1. 237. 5.

Is exceedingly rare in England. Our specimens were taken on Epping Forest in June.

It is suspected that this species, although now taken in England, was not originally a native, but by accident has been introduced into this country, from Germany, or some other part of Europe.

An ingenious collector * informs us, that those taken at Epping are generally found exactly in the same place, and it is worthy a remark, on the same spot there are three posts of foreign fir, which evidently

----

* Mr. Bentley.

L   harbour

harbour a quantity of Larvæ; probably of this infect, though not yet determined.

Has been taken in different parts of the kingdom, and appears to be naturalized with us at this time.

F I G. II. III.

CERAMBYX HISPIDUS.

COLEOPTERA.

CERAMBYX.

SPECIFIC CHARACTER.

Head and thorax fpined, brown. Shells, upper half white with cinereous clouds; lower, brown, with longitudinal ridges, and three ftrong fpines on each, next the future. Antennæ longer than the body, black and white alternately.

> *Linn. Syft. Nat.* 2. 627. 30.—*Fn. Sv.* 651.
> *Geoff. Inf.* 1. 206. 9.—*Fab. Spec. Inf.* 1. 215. 27.
> Cerambyx fafciculatus. *Degeer Inf.* 5. 71. 9. *Tab.* 3. *Fig.* 17.
> Scarabæus. Antennis articulatis longis. *Raj. Inf.* 97. 4.
> *Schaeff. Icon. Tab.* 14. *Fig.* 9.
> *Frifch. Inf.* 13. *p.* 22. *Tab.* 16.

One of the moft beautiful of our Coleopterous Infects, and is common in certain fituations during moft part of the fummer.

Fig. II. reprefents it of the natural fize. Fig. III. magnified.

6

P L A T E

1

2

3

4

PLATE LXV.

FIG. I.

PHALÆNA INTERROGATIONANA,

LEPIDOPTERA.

GENERIC CHARACTER,

Antennæ taper from the base. Wings in general contracted when at rest. Fly by night.

TORTRIX.

SPECIFIC CHARACTER.

Superior wings dark red-brown with an undulated line resembling the note of interrogation on each. Inferior wings and body pale brown.

Is very rare, and has only been hitherto taken in the wilds of Kent, and some other distant parts of the country; our specimen was taken in August.

An insect so singularly marked, cannot readily be confounded with any other species, as we do not possess one which bears much resemblance to it; the most striking particular for a specific distinction are the two waved lines of white on the superior wings, which being contrasted with the brown colour, gives it a very unusual appearance.

It is an undescribed insect, and we have called it Phalæna Interrogationana, as the white undulated mark, if viewed sideways, resembles a note of interrogation.

FIG

PLATE LXV.

FIG. II. III.

PHALÆNA SEMI-ARGENTELLA.

LEPIDOPTERA.

TINEA.

SPECIFIC CHARACTER.

Superior wings gold, with ſtripes of ſilver, inferior wings grey-brown.

―――――――――――

Fig. II. natural ſize.   Fig. III. magnified appearance.

Pha. Semi-argentella is without exception one of the moſt brilliant little moths we have; the natural ſize is ſcarcely ſufficient to diſplay its ſuperior elegance, but when examined by the microſcope, imagination cannot paint a more reſplendent object, for we inſtantly diſcover a moſt wonderful combination of all the varied ſhades of molten ſilver and burniſhed gold; its ſuperior wings are entirely adorned with plates which exhibit in one view the appearance of thoſe coſtly metals, but vary with every direction of light; that which appears gold in one point of ſight becoming red, or bright orange, while the ſhades which were before of a dark brown, aſſumes the reſplendence of burniſhed gold; the thorax glitters with the ſame ſplendor; the head, antennæ, and even the legs, partake alſo of this rich colouring in ſome changes of light; the inferior wings are of a very delicate texture, grey colour, changeable, and though comparatively ſmall, are ſurrounded by a deep fringe, which gives them the appearance of proportion.

We are unacquainted with the works of any author that contain a figure of this inſect, or we might perhaps be enabled to determine
                                                                    whether

PLATE LXV. 77

whether it is not the *Pha. T. Seppella* * of Fabricius; the descriptions nearly correspond, but we are unwilling, without other proof, to give it that specific name.

Until very lately it was considered as an exceedingly rare insect, but several specimens were taken at Highgate last summer.

FIG. IV.

PHALÆNA CURTISELLA.

LEPIDOPTERA.

TINEA.

SPECIFIC CHARACTER.

Superior wings, and thorax white, speckled, and spotted with brown. Inferior wings and body pale brown.

This insect is very uncommon, and though it has never been either figured or described before, it has been arranged in those cabinets which possessed the specimen, under the specific name *Curtisella*, after Mr. CURTIS, author of the *Flora Londinensis*, &c.

The name was originally inserted by Mr. MARSHAM, in his manuscripts, and was intended as a compliment to the abilities of that scientific gentleman; it has not hitherto appeared in public, but we can feel no reluctance to adopt the same name.

---

* Alis auratis, strigis duabus argenteis. *Gen. Inf. Mant.* 296.

PLATE

PLATE LXVI.

BOMBYLIUS MAJOR.

HUMBLE-BEE FLY.

DIPTERA.

Wings two.

GENERIC CHARACTER.

Trunk taper, very long, sharp, between two horizontal valves.

SPECIFIC CHARACTER.

Body short, thick, covered with thick yellowish down. Wings dark brown next the anterior margin; transparent next the posterior margin. Legs long, slender, black.

Linn. Syst. Nat. 2. 1009. 1.—*Fn. Sv.* 1918.
Bombylius variegatus, &c.
   Degeer. Inf. 6. 268. 1. *Tab.* 15. *Fig.* 10.
 Asilus, &c. *Geoff. Inf.* 2. 466. 1.
   Reaum. Inf. 4. *Tab.* 8. *Fig.* 11, 12, 13.
   Mouff. Inf. 64. *Fig.* 5.
   Scop. Carn. 1018.
   Raj. Inf. 273.
   Schaeff. Icon. Tab. 79. *Fig.* 5.
   Huffnag. Inf. Tab. 8. *Fig.* 1.
   Aldr. Inf. 350. *f.* 10.

We

We have only three species of this genus in England, *Major*, *Medius*, and *Minor*.

B. Major is not very rare, its usual time of appearance is June and July.

Together with other species of the Bombylius genus, it is sometimes called the Sword-Bee-Fly: this appellation it receives from the singular form of its trunk; to assist our description, we have represented its appearance when magnified, at Fig. I.

It hovers from flower to flower, when the warmth of the sun invites it abroad, and extracts the nectar from flowers, by darting its proboscis into them, but never rests while feeding.

PLATE

PLATE LXVII.

MELOE VARIEGATUS,

SCARCE MELOE.

COLEOPTERA.

GENERIC CHARACTER.

Antennæ globular, the laft globule oval. Thorax roundifh. Shells foft. Head gibbous, and bent downwards.

SPECIFIC CHARACTER.

Head and thorax dull green, margined with red. Shells fhort, dull green fhagreened. Body large; above variegated with red, green, and copper colour: beneath purple. Legs reddifh purple.

In form and fize this fpecies is not unlike the common Meloe *; but is far fuperior to that Infect, for the beauty of its colours: when the creature is alive, the upper part of the body partakes of the moft vivid colours, but thofe colours become more obfcure after the Infect dies;—this difference of the appearance, between the living and dead fpecimen of the fame fpecies, is not peculiar to this Infect only, but is commonly obferved of moft other kinds. The body is large in proportion to the other parts, but after death it is fo contracted, or diftorted from its natural fhape, as to affume the appearance of an incoherent mafs; the fkin fo corrugated as to receive a falfe light on different parts of the furface, and confequently the natural glow of

* M. Profcarabæus.

M

the

the colours confiderably decreafed by the exhalation of that moifture which ferved to refrefh them in the living ftate.

The underfide, from the greater tenacity of the fkin, or fhelly fub-ftance, is lefs liable to alteration than the upper fide; it is entirely of a dark, but beautiful purple, which is changeable in proportion to the convexity of the body, to the moft brilliant hues; the legs are alfo of a beautiful purple, with the appearance of bronze or copper colour intermixed.

It does not appear to be frequent in any part of Europe; even in Germany it is rarely, if ever taken: as one of the *Britifh Coleopteræ* it is very little known, and is perhaps confined to the diftant parts of Kent, where it is not generally diffufed, but is found local to certain fituations.

Mr. Crow, of Feverfham, very fortunately met with a brood of them laft feafon, and tranfmitted feveral fpecimens to his friends in London. They varied confiderably in feveral refpects, and particularly in their colours; fome appearing much more beautiful than others.

The male is fmaller than the female; they fecrete themfelves be-neath the furface of the earth, and fubfift on the roots of grafs, or herbage in general: are fometimes found by turning up the mould, or may be obferved crawling among the grafs. Come forth in April, or May.

PLATE

69

1

5

3

2

4

PLATE LXVIII.

FIG. I. II.

DYTISCUS MINUTUS.

COLEOPTERA.

GENERIC CHARACTER.

Antennæ taper, or clavato perfoliated.   Feet villous and broad.

SPECIFIC CHARACTER.

Yellow-brown, Shells ftriated, and marked with fhort longitudinal ftripes of black.

Fabri. Spec. Inf. 1. 297. 36.
Chryfomela Minuta.   *Linn. Syft. Nat.*
Dytifcus Minutus.   *Linn. Syft. Nat.* 2. 667. 23.—*Fn. Sv.* 778.
Dytifcus Ruficollis.   *Degeer. Inf.* 4. 404. 18. *Tab.* 16. *Fig.* 9.

Linnæus placed this Infect among the CHRYSOMELÆ, under the fpecific name *Minuta*; but Degeer configned it to the DYTISCUS genus, and gave it the name *Ruficollis:* As a DYTISCUS it alfo appeared in the *Syftema Naturæ*; and Fabricius, as well as other late Entomologifts, have determined it to that genus, either calling it *Minutus*, or after Degeer, *Ruficollis*.

At Fig. I. is reprefented its appearance when magnified, and at Fig. II. the natural fize.

M 2

Is

Is not common; our specimens were taken on *Epping Forest* in *June*. It is an aquatic Insect, or one of that kind which passes through the several states in the water, and subsists on the smaller kinds of Insects, or on the fragments of macerated vegetables. Swims very swiftly.

FIG. III. IV.

DYTISCUS FERRUGINEUS.

COLEOPTERA.

DYTISCUS.

SPECIFIC CHARACTER.

.Very convex. Above red-brown. Beneath paler.
Lin. Syst. Nat.

FIG. III. natural Size, FIG. IV. magnified.

This Insect is one of the same family, and was found at the same time and place as the preceding species. Is not very frequently met with.

F I G.

PLATE LXVIII. 85

FIG. V.

DYTISCUS SULCATUS,

COLEOPTERA.

DYTISCUS.

SPECIFIC CHARACTER.

Shells brown, with four broad furrows, in which are grey-brown hairs. Head black, anterior part yellow, with tranfverfe ftripes. Thorax black, with yellow marks. Beneath black.

Syft. Ent. 231. 6.
Linn. Syft. Nat. 2. 666. 13.—*Fn. Sv.* 773.
Geoff. Inf. 1. 189. 5.
DYTISCUS *fafciatus*, &c. *Degeer Inf.* 4. 397. 4.
HYDROCANTHARIS. *Raj. Inf.* 94. 3. 10.
Frifch. Inf. 13. *p.* 13. *Tab.* 7.
Roes. Inf. 2. *Aquat.* 1. *Tab.* 3. *Fig.* 7.
Bradl. Nat. Tab. 26. *Fig.* 2. A.
Schaeff. Icon. Tab. 3 *Fig.* 3.
Bergftr. Nomencl. 1. *Tab.* 5. *Fig.* 3. 4. 5. *Tab.* 7. *Fig.* 6. 7.

It is fufpected that the DYTISCUS *Sulcatus* is only the female of the DYTISCUS *Cinereus*, and by no means a diftinct fpecies, although Linnæus confidered it as fuch.

It is common in the month of *May*, and thence is found throughout the Summer. It paffes through the different changes, and exifts in the adult ftate in the water; and like others of the fame tribe, devours the fmaller kinds of aquatic Infect, or tender vegetables. It darts with aftonifhing fwiftnefs in fearch of its prey by the affiftance of its hinder legs, which are well contrived for that purpofe.

PLATE

63

PLATE LXIX.

PHALÆNA RUBI.

Fox-coloured Moth.

Lepidoptera.

GENERIC CHARACTER.

Antennæ taper from the bafe. Wings in general contracted when at reft. Fly by night.

No Trunk. Firft Wings horizontal. Second erect.

SPECIFIC CHARACTER.

Antennæ feathered. Wings entire, with a whitifh margin; two whitifh tranfverfe waves on the firft pair.

> *Syft. Ent.* 565. 35.
> *Linn. Syft. Nat.* 2. 813. 21.—*Fn. Sv.* 1103.
> *Wilk. Pap.* 25. *Tab.* 3. *a.* 19.
> *Ammiral. Inf.* 32.
> *Roes. Inf.* 3. *Tab.* 49.

The females of this fpecies are very rarely met with, as they conceal themfelves among the grafs; but the males are commonly taken when flying, and generally indicate that the females are near.

The Caterpillars will feed on the willow, but prefer the leaves of the bramble.

In

In this ftate they are found about the latter end of June, July, or Auguft; and remain fo during the Winter. In April they change to the Pupa form, and in May they appear in the Fly ftate.

The Moth has little to recommend it to notice; and the Pupa, like moft others, is of a dull uniform black brown; it is therefore under the form of a caterpillar that it appears to moft advantage.

P L A T E

1

2 |

| *

2

3

|

PLATE LXX.

FIG. I.

SCARABÆUS TESTUDINARIUS.

COLEOPTERA.

GENERIC CHARACTER.

Antennæ clavated, their extremities fiffile. Five joints in each foot.

SPECIFIC CHARACTER.

Head black without tubercles. Thorax black, punctured, and covered with fhort foft hairs. Shells deeply and equally ftriated, fo as to produce even and regular ridges between the ftriæ, which are of an obfcure black, fprinkled with fmall fpots of a deep yellow. Feet are of a dirty brown colour.

This beautiful animal was defcribed by *Fabricius* as an Englifh Infect in his firft work, the *Syftema Entomologiæ*, but we have never feen a fpecimen of it before. A figure of this Infect may be found in *Fuefly*, *Jablonfky*, and *Olivier*; but thefe works being in few hands, we truft our figure will not be unacceptable to the Englifh Entomologift.

Fig. I. The natural fize denoted by a line.
Fig. I. The magnified appearance.

N   FIG.

FIG. II.

SCARABÆUS CONFLAGRATUS.

Coleoptera.

Scarabæus.

SPECIFIC CHARACTER.

The whole body black and fhining, except the fhells, which are teftaceus, ftriated, with an oblong fpot, rather obfcure on each fide near the external margin. On the head are three tubercles, the middle one larger than the others. Thorax convex and pointed.

This Infect refembles much the *Scarabæus Confpurcatus*, but is a little bigger.

It is alfo figured by *Jablonfky* and *Olivier*, and is defcribed by *Fabricius* in his new Work the Entomologiæ Syftema.

Fig. II. The line fhews the natural fize.
Fig. II. Magnified appearance.

FIG.

PLATE LXX. 91

FIG. III.

SCARABÆUS QUADRIMACULATUS.

COLEOPTERA.

SCARABÆUS.

SPECIFIC CHARACTER.

Head black, without tubercles, but has two little protuberances over the mouth. Thorax black, shining, convex, and covered with impreſſed points. Shells black, ſtriated, with two red ſpots on each, one ſmall at the baſe near the outer margin, the other larger near the apex. Underſide, feet, and antennæ are black and poliſhed.

We are of opinion that the three Inſects in the annexed plate will be new to moſt of our Engliſh Collectors, notwithſtanding they are to be found in this country.

As it would be very difficult, if not impoſſible, to give a juſt repreſentation of theſe minute Inſects in the natural ſize, we have preferred giving the magnified appearance; the outlines which accompany each, and bear the ſame numbers, denote the true ſize of the original ſpecimens.

Fig. III. The line ſhews the natural ſize.
Fig. III. Magnified appearance.

This ſpecies is deſcribed by *Linnæus*, *Fabricius*, and other authors, and has been figured by *Olivier* and *Jablonſky*, being frequently met with in foreign cabinets. It is the ſmalleſt of this genus.

Olivier deſcribes this inſect as having the antennæ and feet red; but it is not ſo in our ſpecimen.

PLATE

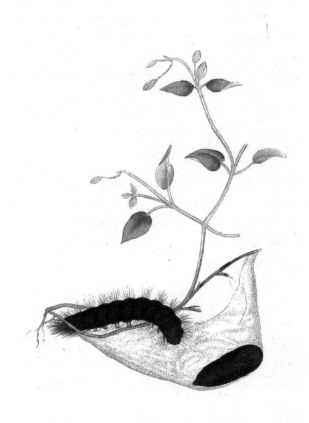

71

PLATE LXXI.

PHALÆNA VILLICA.

CREAM-SPOT TYGER MOTH.

LEPIDOPTERA.

GENERIC CHARACTER.

Antennæ taper from the bafe. Wings in general contracted when at reft. Fly by night.

* No Trunk. Wings depreffed, deflexed. Back fmooth.

SPECIFIC CHARACTER.

Antennæ, head, and thorax black, with a white fpot on each fide the latter. Firft wings black, with eight large cream-coloured fpots. Second wings and body orange, with black fpots.

Syft. Ent. 2. 581. 85.
Linn. Syft. Nat. 2. 820. 41.
Geoff. Inf. 2. 106. 1.
Harris. Aurel. Tab. 4.
Raj. Inf. 156. 4.
Alb. Inf. Tab. 21.
Frifch. Inf. 10. *Tab.* 2.
Reaum. Inf. 1. *Tab.* 31. *Fig.* 4. 6.
Roes. Inf. 4. *Tab.* 28. *Fig.* 2.
———————— *Tab.* 29. *Fig.* 1. 4.
Wilk. Pap. Tab. 3. *a.* 2.

Chickweed is a favorite food with the Caterpillars of this Infect, but it will eat the leaves of the currant, white-thorn, nettle, grafs, &c. if the former cannot be readily procured.

The

The Caterpillars are black and foxy, or hairy; but in a lefs degree than the Caterpillars of Ph. Caja, Great Tyger Moth, which we have figured in the early part of this work.

About the latter end of April the Caterpillars have attained their full fize, and change into chryfalis; late in May they appear in the winged ftate.

It is by no means fo frequent as the Great Tyger Moth, though not very rare; but it is infinitely fuperior for the happy combination of its colours to it, or either of the Britifh fpecies of that tribe which are trivially termed Tygers: it is already high in the efteem of collectors; and were fpecimens of the kind lefs common, it would be in great requeft among the Englifh Entomologifts.

Frequents banks which face the rifing fun.

PLATE

72

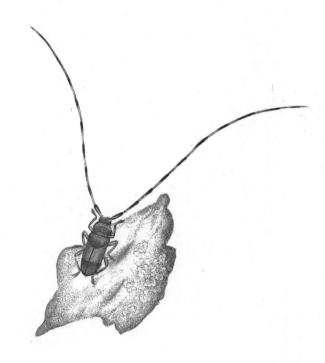

PLATE LXXII.

CERAMBYX ÆDILIS.

Long-horned Cerambyx.

Coleoptera.

GENERIC CHARACTER.

Antennæ articulated, and tapering to the end. Shells long and narrow. Four joints on each foot. Thorax, with lateral fpines or tubercles.

SPECIFIC CHARACTER.

Antennæ confiderably longer than the body. Head, thorax, and fhells grey, with fhades of brown, fprinkled with yellow, and dark brown fpots. Thorax fpined.

> *Syft. Ent.* 164. 1.—*Linn. Syft. Nat.* 2. 628. 37.—
> *Fn. Sv.* 653.
> Cerambyx, &c. *Linn. It. Oel.* 8.
> *Degeer. Inf.* 5. 66. 5. *Tab.* 4. *Fig.* 1. 2.
> Capricornus rufficus. *Petiv. Gazoph. Tab.* 8. *Fig.* 8.
> *Mouff. Inf.* 151. *Fig.* 2.
> *Frifch. Inf.* 13. *Tab.* 12.
> *Sulz. Hift. Inf. Tab.* 4. *Fig.* 27.
> *Act. Nidrof.* 4. *Tab.* 16. *Fig.* 8.
> *Schaeff. Icon. Tab.* 14. *Fig.* 7.
> *Bergftr. Nomencl.* 1. 3. 5. *Tab.* 1. *Fig.* 5. 6.
> *Tab.* 2. *Fig.* 1.
> *Fab. Spec. Inf.* 1. 209. 1.

This fpecies is found in every part of Europe, though very fcarce; and in England it is extremely rare.

And

And it is no lefs diftinguifhed for the very fingular ftructure and length of its antennæ, than for its rarity; that part which forms one of the moft certain characteriftics of almoft every tribe of Infects, conftitutes the moft prominent character in this.

Of its ufe, we are altogether ignorant, as the various opinions that have been given by former writers are now obliterated; fome have fuppofed that they were the organs of hearing, or fmell; and others have imagined that they were fufceptible of the leaft motion in the ambient fluid in which they move.

Geoffroy difcovered the organs of hearing in feveral amphibious animals, viz. in the toad, frog, viper, fome other ferpents, lizard, water-falamander, and fkate *; and many of the moft eminent ana-tomifts of the prefent time have difcovered by their refearches into the animal kingdom, thofe organs in different creatures. Profef-for Camper, in 1763, publifhed remarks on the organs of hearing in fifhes, in the Harlem Tranfactions †: Mr. Hunter has defcribed others in the Philofophical Tranfactions ‡; and Dr. Monro has de-fcribed and figured great variety of them in his large work on the ftructure and phyfiology of fifhes.

Probably, induced by thofe difcoveries profeffor Fabricius endea-voured to afcertain the organs of hearing in Infects alfo; and about nine years ago publifhed an account of this interefting difcovery in the New Copenhagen Tranfactions §, with figures of thofe organs in the crab and lobfter: he found the external orifice of the organ in thefe animals to be placed between the long and the fhort antennæ, the cochlea, &c. being lodged in the upper part, which Linnæus calls the thorax, near the bafe of the ferrated projection at its apex; we muft therefore conclude that the antennæ of Infects are appropriated for fome other purpofes than thofe it is at prefent fufpected they anfwer.

The Cerambyx Ædilis, Fabricius informs us, lives in the trunks of trees; its horns are moveable, as it can either direct them forward, or fupport them in an erect pofition; and when it fleeps, it reclines them along its back; it alfo reclines them when it walks quick, or has to pafs through a narrow track, as the leaft refiftance from any thing in its way, would be very liable to injure, or break them off.

Our fpecimen was taken in May.

---

* Memoires Etrangers de l'Acad. de Paris, 1755.
† In the Year 1763, &c.   ‡ Vol. lxxii.   § Vol. ii. p. 375.

LINNÆAN INDEX

TO

VOL. II.

COLEOPTERA.

The Star * diſtinguiſhes thoſe which have not been named before.

HEMIP.

INDEX.

HEMIPTERA.

LEPIDOPTERA.

NEU-

I N D E X.

NEUROPTERA.

HYMENOPTERA.

DIPTERA.

APTERA.

ALPHABETICAL INDEX

TO

VOL. II.

INDEX.

ERRATA

ERRATA to Vol. II.

Page 11, for Plate XL. *read* XLI.

Plate LXVIII, fhould have been numbered LXIX.

Plate LXIX, fhould have been numbered LXVIII.

THE

NATURAL HISTORY

OF

BRITISH INSECTS;

EXPLAINING THEM

IN THEIR SEVERAL STATES,

WITH THE PERIODS OF THEIR TRANSFORMATIONS,
THEIR FOOD, ŒCONOMY, &c.

TOGETHER WITH THE

HISTORY OF SUCH MINUTE INSECTS

AS REQUIRE INVESTIGATION BY THE MICROSCOPE.

THE WHOLE ILLUSTRATED BY

COLOURED FIGURES,

DESIGNED AND EXECUTED FROM LIVING SPECIMENS.

By E. DONOVAN.

VOL. III.

LONDON:

PRINTED FOR THE AUTHOR,

And for F. and C. RIVINGTON, Nº 62, ST. PAUL's CHURCH-YARD.

MDCCXCIV.

S

PLATE LXXIII.

PAPILIO LATHONIA.

LESS SILVER-SPOTTED BUTTERFLY,

OR,

QUEEN OF SPAIN.

FRITILLARY.

LEPIDOPTERA.

GENERIC CHARACTER.

Antennæ clavated. Wings, when at reft, erect. Fly by day.

SPECIFIC CHARACTER.

Above yellow-brown, with fpots of black. Beneath yellowifh, variegated with dark brown, and black fpots. Thirty-feven filver fpots on the pofterior wings.

> *Syft. Ent.* 5. 17. 314.—*Linn. Syft. Nat.* 2. 786. 213.—
> *Fn. Sv.* 1068.—*Geoff. Inf.* 2. 120. 6.—*Fab. Sp.*
> *Inf.* 2. 110. 481.

Papilio Rigenfis minor aureus, maculis argenteis fubtus perbelle notatus.—*Raj. Inf.* 120. 6.

> *Hufn. Pict. Tab.* 11. *Fig.* 11.
> *Robert. Icon. Tab.* 12.
> *Merian. Europ.* 2. *Tab.* 157.
> *Roes. Inf.* 3. *Tab.* 10.
> *Efp. Pap.* 1. *Tab.* 18. *Fig.* 2.
> *Schaeff. Icon. Tab.* 143. *Fig.* 1. 2.
> *Seb. Muf.* 4. *Tab.* 1. *H.* 1—4.

B   We

We have feveral fpecies of the Papilio tribe, which are highly valued in England either for their beauty or fcarcity; the *P. La-thonia* is little, if by any means, inferior to the moft beautiful; and as a rare Infect is efteemed an invaluable acquifition.

The upper fide is only a plain orange or brown colour, with fpots of ftrong black, and does not in general appearance differ materially from the greafy Fritillary Butterfly, which is very common in moft fituations; but the underfide is entirely unlike every other Englifh Infect: the bright filver fplafhes on the under wings are fingular in their form, and fo beautifully relieved by the orange ground colour, and variegation of black between, as to form a delightful contraft of the moft pleafing colours.

Whether this fpecies was originally a native of this country, may be doubtful; we certainly have the moft refpectable teftimonies of its being taken alive in different parts of the kingdom, but it might have been introduced by accident in the larva, or more probably in the pupa ftate: it has been feen at *Bath*; and either *Mofes Harris*, or fome of his friends, bred it from the caterpillar. " Queen of Spain Fritillaria changed into chryfalis *April*, appeared in the winged ftate *May* 10th.—*Gambling Gay wood*, near *Cambridge*."

It has alfo been taken near *London*; Mr. *Honey*, of *Union-ftreet*, in the *Borough*, took one a few years fince in his garden. I requefted the favour of whatever information he could communicate refpecting this circumftance, and received a note with thefe words:—

" *September* 9th, 1785.—I took the Queen of Spain Butterfly in my garden.   (Signed)   WM. HONEY."

P L A T E

PLATE LXXIV.

CURCULIO BETULÆ.

COLEOPTERA.

GENERIC CHARACTER.

Antennæ clavated; elbowed in the middle, and fixed in the fnout, which is prominent and horny. Joints in each foot, four.

* Snout long.

SPECIFIC CHARACTER.

Green-gold. Antennæ and eyes black. The anterior verge of the thorax fpinous in one fex only.

> *Syft. Ent.* 130. 16.
> *Linn. Syft. Nat.* 2. 611. 39.
> *Fn. Sv.* 605.
> *Degeer Inf.* 5. 248. 5. *Tab.* 7. *Fig.* 25.
> Rhinomacer, &c. *Geoff. Inf.* 1. 270. 2.
> *Frifch. Inf.* 12. 17. *Tab.* 8. *Fig.* 2.
> *Sulz. Hift. Inf. Tab.* 4. *Fig.* 5.
> *Schaeff. Icon. Tab.* 6. *Fig.* 4.

The Linnæan defcription of the *Curculio Betulæ,* fo nearly correfponds with that of *C. Populi,* that if we allow for the variation of colour to which all Infects are fubject, a line can fcarcely be drawn between the two fpecies; the moft material diftinction is the underfide of *C. Betulæ* being of the fame colour as the back; but the underfide of *C. Populi* is purple, and fmaller.

B 2

The

The defcription which *Linnæus* has given of our fpecies is, " longi-
" roftris, corpore viridi aurato fubtus concolore ;" and *Degeer* has defcribed it in fimilar words. It is evident that *Linnæus* had reafon to
fufpect fome difference of colour between the two fexes, but perhaps
he never imagined the *C Purpureus* *, which he had before defcribed,
was alfo one fex, or a variety of the fame Infect.

Geoffroy fays, " *Rhinomacer* totus viridi cœruleus ;" and *Fabricius*
adds, " Variat fæpius colore omnino cœruleo. Alter fexus thoracem
" antrorfum fpinofum gerit." How thofe different defcriptions may
be reconciled, fo as to be defcriptive of the two fexes of *C. Betulæ*,
will appear more clearly on farther obfervation.

Late in *May*, this feafon, being at *Darent-Wood*, *Dartford*, I met
with one of the green kind, and one of a dark blue colour, with a
fhade of green on the elytra ; I could not be deceived, they were male
and female ; as a farther corroboration, I met with a fecond pair, in
a fimilar fituation ; and on the day following a third : the blue one of
this laft pair had not the fhade of green as on the former, but was of
a rich gloffy blue purple ; and I am greatly miftaken if it is not the
C. Purpureus of *Linnæus*, or the Infect which is arranged in *Englifh*
Cabinets for that fpecies.

I communicated the circumftance of meeting with thofe two Infects,
which have always been confidered as diftinct kinds, to a perfon who
alfo was collecting Infects in the wood, on the fame day, and he informed me that he had juft before difcovered them in the fame fituation.
I have examined them very carefully, but cannot difcover any fpines
on the thorax of the green and gold kind, though I have five of them,
but the three purple fpecimens are all fpinous, as defcribed by authors.
I am of opinion, that the bright coloured fpecimens are all *females*,
and thofe which are purple, I imagine, are *males*.

I mentioned the circumftance to an eminent Entomologift, and he
at firft fufpected they might be mule Infects, generated between the

---

* *Berkenhout*, in his Outlines of the Natural Hiftory of *Great-Britain*, fays,
C. Purpureus. Gloffy Purple. Snout very long. *Petiver* found this at
Epfom.

C. Betulæ

PLATE LXXIV. 5

C. Betulæ and the *C. Purpureus*, but that could not be the cafe, as they were all in copulation when taken.

We have been the more minute in this account, as we confider the confounding of one fpecies with another fhould ever be avoided, with as much care as the feparation of varieties into diftinct fpecies; both tend to confufe, or fubvert that truth which fhould be the guide of every enquirer into nature.

I have received a letter from my refpectable friend *T. Marfham*, Efq; Sec. L. S. accompanied with a fpecimen of the green kind of *C. Betulæ*, that is fpinous on the thorax; together with one of the blue or purple kind, which is fpinous alfo : he informs me, that though his purple fpecimen has fpines, he is very certain he has had one without; hence it appears to me that they admit of great variation; indeed it would afford the moft prefumptive argument, that there are males and females of both colours. Among the purple fpecimens which I took, there was a confiderable difference in their colours, but of the five green fpecimens fcarcely two exactly agreed; one in particular partook fo much of a vivid crimfon that it might eafily have been miftaken by a curfory collector for the *Curculio Bachus*.

PLATE

PLATE LXXV.

NOTONECTA GLAUCA.

COMMON BOAT-FLY.

HEMIPTERA.

GENERIC CHARACTER.

Antennæ beneath the eyes. Wings croffed, and complicated. Feet formed for fwimming. Hind feet hairy.

SPECIFIC CHARACTER.

Head yellow; eyes brown, large. Thorax, anterior part yellow, pofterior black. Shells pale yellow brown, with a bright brown anterior margin, fpotted with black. Beneath brown. Feet of two joints. Length fix lines.

Syft. Ent. 689. 1.—*Linn. Syft. Nat.* 2. 712. 1.—
Fn. Sv. 903.

Notonecta, &c. *Geoff. Inf.* 1. 476. 1. *Tab.* 9. *Fig.* 6.
Nepa notonecta, &c. *Degeer Inf.* 3. 382. 5. *Tab.* 18. *Fig.* 16. 17.
Cicada aquatica *Mouffeti. Raj. Inf.* 58.
Notonecta, &c. *Petiv. Gazolph. Tab.* 72. *Fig.* 6.
Cimex aquaticus anguftior. *Frifch. Inf.* 6. 28. *Tab.* 13.
Cimex aquaticus. *Roef. Inf.* 3. *Tab.* 27.
Bradl. Nat. Tab. 26. *Fig.* 2. E.
Huffnag. Inf. Tab. 12. *Fig.* 19.
Sulz. Inf. Tab. 10. *Fig.* 67.
Schaeff. Elem. Tab. 90.
—— *Icon. Tab.* 33. *Fig.* 5. 6.
Fuefly Inf. Helv. 24. 468.

This

3

This fpecies is by far the moft common of the Notonecta genus in *England*. It is an aquatic Infect, undergoes its feveral changes in the water, and in the laft ftate is furnifhed with wings for flight.

In the day-time it may be obferved on the furface of ftill waters; it always fwims on its back, with its legs extended. In the evening it flies in the air. Found during moft of the fummer months. Moft authors have defcribed the upper fhells as being of a brown colour, variegated with clouds of black; but this appearance is not conftant in every fpecimen; they lofe much of that colour after being taken out of the water, or if the wings are expanded.

P L A T E

PLATE LXXVI.

PHALÆNA FUNALIS,

FESTOON MOTH.

LEPIDOPTERA.

GENERIC CHARACTER.

Antennæ taper from the bafe. Wings, when at reft, generally contracted. Fly by night.

SPECIFIC CHARACTER.

Upper wings orange, rather inclining to brown; with a black line nearly of a triangular form on each; when the wings are expanded the lines refemble a feftoon. Under wings orange, clouded and frofted with black; margin pale.

We are happy to prefent our Subfcribers with the figure of a Moth which is fcarcely known among the Englifh Collectors, and we may venture to affert on the beft authority has not a place in any cabinet of Infects in this metropolis, except that of the Author; indeed the only perfon who appears to have been fo fortunate as to meet with it except himfelf, is Mr. Lewin, who formerly refided at Dartford; he confidered it as fuch an invaluable rarity, that had not a figure of it been difcovered in *Roefel*, it would no doubt have been publifhed in the Tranfactions of the Linnæan Society; it muft, however, be obferved, that the Infect *Roefel* has figured is a foreign fpecimen.

C

On

On the communication of Mr. Jones, of *Chelsea*, we presume that
this Insect was formerly known among the English Collectors, and
received from them the appellation *Festoon Moth*, but it must have been
extremely rare even at that time, as it does not appear in Harris's List
of English Moths, nor has a single specimen, or its remains, been
found among the old Collections, which have been handed down to
the Entomologists of the present day.

On the 16th of August, 1793, I shook the Caterpillar from one of
the high branches of an oak-tree, in Darent wood, Kent; it remained
motionless for some time when in the net, and I concluded that it might
have sustained an injury by its fall; but I soon after discovered that it
was naturally a sluggish, inactive creature, and had received no damage; it remained several days in the Caterpillar state, but as it was
almost ready to change into Chrysalis, I had only an opportunity of
being convinced that oak was its proper food.

This Caterpillar is a most singular creature; at one time it would
flatten itself, and be considerably extended in breadth, or length; at
another time it would gather itself up like an hedge-hog, or become
almost round, and in a few minutes after it would be flat again; and
frequently the orange colour on the back would be obliterated; sometimes it so nearly resembled the Caterpillars of several of the Papilio
tribe, that I suspected it to be one of the *Hair-streak Butterflies*, or
rather the Caterpillar of a new species. On the 23d of August it
began to spin, and in a short time after its case was completed.

The case in which it passed to the Pupa state, was very firmly con-
structed, and precluded an opportunity of observing the different
symptoms of change, which would otherwise have been visible. This
case, which was exactly in the form of an egg, was at first of a pale
flesh colour, but in the course of a few days it had heightened to a
very fine sanguineous, and after to a scarlet, or nearly vermilion colour; this colour it retained for several months, but as the time for
the emancipation of the Moth within approached, the brightness of
                                                                    red

red fomewhat abated, though even after the Fly came forth, much of the original colour remained.

The manner in which it burfts open the cafe is rather fingular; it does not force an opening in an irregular form, as moft Infects which fpin a cafe, but defcribes an exact circle within at one end; after this it divides its cafe according to that circle, only leaving a fmall portion to act as an hinge; when it has extricated itfelf from the Chryfalis, it forces the top of the cafe back, as fhown in our Figure, and thereby a free paffage is opened for its delivery.

The infide of the cafe is perfectly fmooth, and appears as if po-liſhed by art; it is of a pale blue colour, the Chryfalis within is brown.

The Fly came forth on the 12th of July, 1794.

PLATE

2

3

1

4

P L A T E LXXVII.

FIG. I. and FIG. II.

PHALÆNA CRISTALANA.

DARK-BUTTON MOTH.

LEPIDOPTERA.

GENERIC CHARACTER.

Antennæ taper from the bafe. Wings, in general contracted when at reft. Fly by night.

TORTRIX.

SPECIFIC CHARACTER.

Upper wings yellow-brown, with dark fhades; a broad irregular white mark, and a tuft or button, on the center of each. Head and thorax white clouded. Lower wings pale brown.

———————

This fingular Moth is very rarely met with; it has been taken in *Coombe-wood, Surry,* and in *Kent,* but even in thofe places it is very uncommon.

It is diftinguifhed by the unufual form of the white markings on the upper wings, and particularly by the tuft or button which is fituated in an upright pofition near the center of each; thofe tufts appear only flightly feathered on the upper parts to the naked eye, but when one of them is examined with a microfcope, or even common magnifier, it prefents the appearance of a bundle of fibres, inclofed within a thin membrane; narrow at the bafe, encreafing in bulk near the middle,

2   and

and expanding at the fummit into a number of fhoots, in the form of a creft: feveral other tufts are difperfed near the extremities of the upper wings, but they are not confpicuous to the naked eye.

I have feen an Infect which correfponds in every refpect with this fpecimen, except that it had a line of a dull ochre colour along the pofterior margins of the upper wings; but I fufpect it to be either a variety, or perhaps only the difference of fex.

Linnæus has not defcribed this infect, neither can we difcover any defcription of it in the writings of *Fabricius*; and I have no doubt of its being a nondefcript fpecies.

The fingular crefted tufts, with the white markings on the upper wings, furnifh fuch an ample fpecific diftinction, that we have named it *Criftalana*.

Found early in the month of *Auguft*.

Fig. I. reprefents the natural fize. Fig. II. its magnified appearance.

F I G. III. and F I G. IV.

P H A L Æ N A   R A D I A T E L L A.

LEPIDOPTERA.

PHALÆNA.

TINEA.

SPECIFIC CHARACTER.

Firft wings, buff, with fhades of orange; ftriped or rayed with a very dark purple from the bafe to the apex of each; a white ftripe near to, and parallel with the pofterior margin, and two fpots of the fame colour near the center of each wing. Second wings lead colour; deeply fringed.

This

PLATE LXXVII. 15

This infect alfo appears to be a nondefcript fpecies; we have called it *Radiatella*, or rayed, from the form of the dark ftripes which rife from the bafe, and fpread in the form of rays to the apices of the upper wings. It is very liable to change after death, and particularly the buff colour, which appears very bright when the infect is frefh, but is fometimes fo altered in appearance when placed in the cabinet, that an intermixture of that colour can be fcarcely diftinguifhed between the rays of purple; we mention this circumftance, as very few fmall lepidopterous infects are fubject to fuch alteration.

It is found about the fame time as the Phalæna Criftalana, and I believe is equally fcarce.

Fig. III. reprefents the natural fize. Fig. IV. its magnified appearance.

PLATE

1

4

3

2

PLATE LXXVIII.

FIG. I. and FIG. II.

CHRYSOMELA BOLETI.

COLEOPTERA.

GENERIC CHARACTER.

Antennæ knotted, enlarging towards the ends. Corfelet margined.

SPECIFIC CHARACTER.

Antennæ, head, and thorax black, fhining. Elytra black, with two jagged belts of bright orange colour; extremity orange.
Syft. Ent. 97. 18.—*Linn. Syft. Nat.* 2. 591. 36. —*Fn. Sv.* 52. 7.—*Sulz. Hift. Inf. Tab.* 3. *Fig.* 9.

Diaperis, *Geoff. Inf.* 1. 337. *Tab.* 6. *Fig.* 3. *mal.*
Diaperis, *Schaeff. Elem. Tab.* 58.—*Icon. Tab.* 77. *Fig.* 6.
Dermeftes, &c. *Vdm. Diff.* 4. *Fig.* 3.
Tenebrio Boleti, &c. *Degeer Inf.* 5. 49. 9. *Tab.* 3. *Fig.* 3.
Coccinella fafciata. *Scop. Ent.* 247.

The Chryfomela Boleti is not very frequent in this country; it is almoft invariably found in the hollows of fome of the Boletus tribe of Fungi *, which grow on the ftumps of trees in the month of *May* or *June.*

---

* *Mufhrooms.*

D

FIG.

F I G. III. and F I G. IV.

CHRYSOMELA CERUINA.

COLEOPTERA.

CHRYSOMELA.

SPECIFIC CHARACTER.

Oblong. Dull brown, befet with very fine hairs.
Syft. Ent. 116. 1.
Linn. Syft. Nat. 2. 602. 115.—*Fn. Sv.* 575.

There can remain very little doubt of thofe infects N° III. and N
IV. being fexes of the fame fpecies.

Rarely met with near *London. May* and *June.*

P L A T E

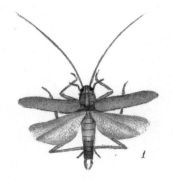

PLATE LXXIX.

GRYLLUS BIGUTTULUS.

HEMIPTERA.

Shells, or upper wings, femi-cruftaceous, not divided by a ftraight future, but incumbent on each other. Beak curved downward.

GENERIC CHARACTER.

Head maxillous, and with palpi. Antennæ filiform, or taper. Wings folded. Hind legs ftrong for leaping.

SPECIFIC CHARACTER.

Head and thorax dark brown, marked with lines of white. Wings pale brown edged with yellow, and feveral whitifh marks near their extremity. Body beneath, and legs, red-brown.
Linn. Syft. Nat. 2. 702. 55.—Fn. Sv. 875.

ACRIDIUM BIGUTTULUM, &c. *Degeer. Inf. 3. 479. 6.*
GRYLLUS BIGUTTULUS. *Schaeff. Icon. Tab. 190: Fig. 1. 2.—*
Fab. Spec. Inf. 1. 370. 45.

Though few infects require more elucidation to be well underftood than thofe of the *Gryllus* genus, no part of the fcience has been lefs regarded even by fyftematic writers, who certainly appear to have been moft interefted to obtain a fatisfactory knowledge of them : the prefent fpecies is continued by *Fabricius*, in his *Species Infectorum*, under the *Linnæan* genus, and fpecific name GRYLLUS BIGUTTULUS.

All of the Grylli are very liable to variations in colour, and particularly after death ; green changes to brown of various hues, the light colours become dark, and the dark colours fade, fo that no juft idea of the true appearance can be formed except from the living infects.

The

The larva, and pupa, of moſt ſpecies of the Gryllus genus, ſcarcely differ in appearance from the perfect inſect, except that in the two firſt ſtates they are apterous, or without wings, and either leap or walk; but in the laſt ſtate they are furniſhed with four membranaceous wings.

The ſubject of our preſent deſcription is not an unfrequent ſpecies near London; it is taken in the perfect ſtate in the month of Auguſt.

L O C U S T A   V A R I A.

S P E C I F I C   C H A R A C T E R.

Antennæ very long. Thorax green, with a longitudinal line of yellow. Anterior wings membranaceous, green. Poſterior wings very delicate pale green. Body pale green, with the three laſt joints pale black.

Syſt. Ent. 287. 24.
Locuſta thalaſſina, &c. *Degeer. Inſ.* 3. 433. 3.
Goed. Inſ. 2. 142. *Tab.* 40.
Friſch. Inſ. 12. *Tab.* 2. *Fig.* 4.
Sulz. Hiſt. Inſ. Tab. 8. *Fig.* 9.
Locuſta Varia, *Fab. Spec. Inſ.* 1. 360. 25.

It is very plenty in the month of Auguſt, is concealed among the foliage of the lower branches of the oak in the day-time, and is not often obſerved to fly except when the morning dew is on the herbage, or evening approaches. Leaps, if diſturbed.

PLATE LXXX.

PHALÆNA FULIGINOSA.

RUBY-TIGER MOTH.

LEPIDOPTERA.

GENERIC CHARACTER.

Antennæ taper from the bafe. Wings, in general, contracted when at reft. Fly by night.

* Spiral trunks; back fmooth without creft.

SPECIFIC CHARACTER.

Superior wings red brown; a black dot near the center of each. Inferior wings, rofe colour with black marks *. Abdomen, rofe colour with a chain of black fpots down the center, and a row of dots on each fide.

> *Syft. Ent.* 588. 111.
> *Linn. Syft. Nat.* 2. 836. 95.—*Fn. Sv.* 1159.
> *Raj. Inf.* 228. 13.
> *Harr. Aurel. Tab.* 12.
> —— *Inf. Anglic. Tab.* 8. *Fig.* 7.
> *Ammir. Inf. Tab.* 30.
> *Roef. Inf.* 1. *Phal.* 2. *Tab.* 43.
> *Wilk. Pap. Tab.* 3. *a.* 14.

---

* The black marks on the under wings of different fpecimens vary very much; in fome the black occupies half the fpace of the wings; in others the rofe colour is predominant.

E   The

The leaves of Alder or Birch, the Turnip, Muſtard, and Rag-
wort, with many other vegetables, are noticed by different authors,
as being proper food for the Ruby Tiger Moth in the larva ſtate ;
I have obſerved that they prefer the leaves of the Ragwort or
Groundſel.

The Caterpillars are ſmall in the month of May, in June they paſs
to the pupa form, and early in the month following, appear in the
winged ſtate *.

This ſpecies is leſs frequent than the Cream Spot Tiger Moth †,
lately figured in this work.

---

* In a forward ſeaſon like the preſent, the time of their appearance in the different
ſtates may vary confiderably, eſpecially as ſome may have two, or even three broods
in one ſummer. I have a Moth from a ſecond brood, which paſſed to the pupa form
the 25th of July, and came forth the 10th of Auguſt, 1794.

† *Ph. Villica.*

P L A T E

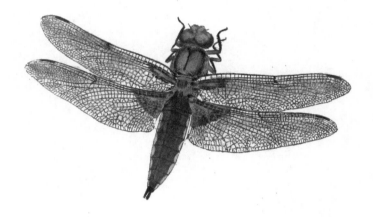

PLATE LXXXI.

LIBELLULA DEPRESSA,

NEUROPTERA.

GENERIC CHARACTER.

Wings four. Naked, tranfparent, reticulated with veins or nerves. Tail without fting.

SPECIFIC CHARACTER.

Eyes brown. Head and thorax greenifh, with two yellow tranf-verfe lines. A dark fpot on the exterior margin of the wings. Body rather depreffed; that of the female, bright brown with yellow marks on each divifion; that of the male, blue grey, with fimilar marks of yellow.

Syft. Ent. 420. 2.
Linn. Syft. Nat. 2. 902. 5.—Fn. Sv. 1413.
Libellula, &c. Geoff. Inf. 2. 226. 9.
Libellula, &c. Raj. Inf. 49. 5.
Reaum. Inf. 6. Tab. 35. Fig. 1.
Roef. Inf. 2. Aqu. Tab. 6. Fig. 4.
Tab. 7. Fig 3.
Edw. Av. Tab. 333.

The Male Infect of the *Libellula Depreffa*, differs fo very mate-rially in colour from the female of that fpecies, that we cannot ima-gine it will be improper to give a figure of the former in our prefent

I   Number,

Number, though the latter is already reprefented in the early part of the Work.

We have nothing particular to add to our former account of its hiftory. In the larva and pupa ftate, it is found crawling at the bottoms of pools or ditches, and fubfifts on the larvæ of Gnats and other Infects; but in the laft ftate, it leaves its aquatic abode, and fubfifts on fmall winged infects, efpecially Moths; it is not uncommon to fee one of this fpecies ftop fhort in its flight, dart down like a Hawk upon a Moth or Butterfly, and tear it to pieces in an inftant; or fly with it in its mouth, to fome more convenient place to devour it.

PLATE

PLATE LXXXII.

PHALÆNA USTULARIA.

EARLY THORN MOTH.

LEPIDOPTERA.

GENERIC CHARACTER.

Antennæ taper from the bafe. Wings in general contracted when at reft. Fly by night.

* Geometræ.

Antennæ of the male feathered; of the female fetaceous, or like a briftle.

SPECIFIC CHARACTER.

Wings angulated, indented, light brown varied with fhades of a fcorched colour. Three waves of dark brown acrofs each fuperior wing; together with a fpot of orange or bright brown colour, at the bafe, and another nearly of the fame colour on the exterior margin of each.

———————————————

Among the feveral Moths of the *Geometræ* divifion of Phalenæ which are known to the Englifh Collectors by the trivial diftinction, *Thorn Moths*, our prefent Infect is neither the moft confpicuous, or rare; it is however a beautiful creature when taken immediately from the Pupa cafe, but rarely fine, when caught in the fly ftate, in the fowling-net; the down being of fuch an exquifite texture that the flighteft touch muft inevitably damage its appearance.

F

The

The Pupa is marked with a brown colour at every annulation immediately after the Caterpillar has paſſed to that ſtate, but as the creature within becomes more perfect, that brown is gradually changed to a dark, or black colour.

I have obſerved much variation in the colours of different ſpecimens of this ſpecies; of three male Inſects which I have bred this ſeaſon, one only correſponded with the annexed figure, one inclined much more to a red brown, and the other to a dull purple.

I met with the Caterpillars on the oak, and they always preferred that food to any other. The Caterpillars are ſmall in July, they paſs to the Pupa ſtate in Autumn, and the Moths are to be taken about the middle of March.

Although, as we have juſt obſerved, this Inſect does not particularly claim our regard as a rarity, it does not appear to have been deſcribed by *Linnæus*, or even by *Fabricius* in his Species Inſectorum; and no account of it is included in Berkenhout's Outlines, in Harris's Catalogue of Engliſh Inſects, or any other work we have had an opportunity of peruſing.

In its manners, the Caterpillar is not more ſingular than in its form; when young it is very active and in continual motion; but as it grows larger it becomes more ſluggiſh in its diſpoſition: it will sometimes affix itſelf by its hind feet to one of the extreme branches of the tree on which it feeds, in the ſame manner as ſhewn in our plate, and will remain in that poſture ſeveral hours without the leaſt apparent ſigns of life.

P L A T E

3

2

PLATE LXXXIII.

CICADA CORNUTA.

HORNED CICADA.

HEMIPTERA.

Shells or upper wings femi-cruftaceous, not divided by a ftraight futu re, but incumbent on each other. Beak curved downward.

GENERIC CHARACTER.

Antennæ taper. Shells membraneous, in each foot three joints. Hind legs ftrong for leaping.

SPECIFIC CHARACTER.

Black-brown. Antennæ fhort. Thorax bicornuted, with the pofterior part elongated almoft to the extremity of the abdomen. Wings diaphanous, croffed. Brown veins on the fhells.

Syft. Ent. 676. 8.

Linn. Syft. Nat. 2. 705.—*Fn. Sv.* 879.

CICADA, &c. *Geoff. Inf.* 2. 243. 18.

Schreb. Inf. 11. *Fig.* 3. 4.

Degeer. Inf 3. 181. 3. *Tab.* 11. *Fig.* 22.

Ranata cornuta. *Petiv. Gozoph. Tab.* 47. *Fig.* 2. 3.

Sulz. Inf. Tab. 10. *Fig.* 63.

Schœff. Icon. Tab. 96. *Fig.* 2.

Scop. Carn. 340.

Membracis cornuta. *Tab. Spec. Inf.* 2. 317. 9.

F 2

The

The Cicada Cornuta is a native of Germany and other parts of Europe, as well as of England; with us it is by no means common. It is met with in the month of May, or June; Berkenhout fays it is found on trees, ferns, &c. I have taken two fpecimens this feafon, one at Coombe-wood, Surrey, the other at Dartford; they were both concealed on the under fides of fome dock leaves.

At Fig. I. the creature is reprefented of the natural fize, with the wings expanded; at Fig. II. one is given in a ftanding pofition; and at Fig. III. the front of the head and fingularly conftructed thorax is fhewn as they appear before the fpeculum of an opaque microfcope.

Fabricius has placed this Infect in a divifion of the feventh Clafs of his Genera Infectorum, RYNGOTA *Membracis.*

PLATE

1

2

3

4

PLATE LXXXIV.

FIG. I.

LEPTURA ARCUATA.

GREAT WASP BEETLE.

COLEOPTERA.

GENERIC CHARACTER.

Antennæ tapering to the end.   Shells narrower at the apex.   Thorax fomewhat cylindrical.

SPECIFIC CHARACTER.

Black.   Antennæ length of the body.   Target yellow.   Three tranfverfe yellow lines on the head; three on the thorax and three yellow arched lines, with as many fpots of the fame colour on each fhell.

LEPTURA ARCUATA.   *Linn. Syft. Nat.* 2. 640. 21. *ed.* XIII.—
    Fn. Sv. 696.

LEPTURA, &c.   *Geoff. Inf.* 1. 212. 10.

CERAMBYX niger, &c.   *Vdm. Diff.* 30.

SCARABÆUS, &c.   *Frifch. Inf.* 12. *Th. n.* 22. *p.* 31. *Tab.* IV.
    Fig. 1—5.

CERAMBYX, &c.   *Leche Nou. Spec.* 30.

SCARABÆUS.   *Raj. Inf.* 83. 23.
    Petiv. Gazoph. Tab. 63. *Fig.* 7.
    Schœff. Icon. Tab. 38. *Fig.* 6.
      Tab. 107. *Fig.* 2. 3.

CALLIDIUM arcuatum.   *Fab. Spec. Ent. n.* 26. *p.* 192.
    Spec. Inf. T. I. n. 35. *p.* 241.
    Mant. Inf. T. I. n. 50. *p.* 155.
    Ent. Syft. T. II. n. 64. *p.* 333.

5   *Der*

Der Bogen-Widderkäfer. Der Bogenstrich. Der Holzkäfer mit Bogenbinden. ·La Lepture aux croissans dorés, Panzer Faun. Inf. Germ. In. N° IV. p. 14.

This is the rarest species of the Leptura genus we have in England; it is found among rotten wood. May.

Fabricius having separated the Lepturæ of Linnæus, and arranged them under three distinct generic divisions, as Callidium, Donacia, and Leptura, it will be proper to observe, that the CALLIDIUM *Arcuatum, Cass I.* ELEVTERA, *Fab. Spec. Inf.* is the LEPTURA *Arcuata* of Linnæus; to this we must also add that the LEPTURA *Arcuata,* figured in the seventh Number of *Panzer's* Faunæ Insectorum Germanicæ Initia, is a very different species to our specimen, is a native of Austria, and received its name from *Hellwig.*

FIG. II.

LEPTURA MYSTICA.

SPECIFIC CHARACTER.

Antennæ and legs black. Head and thorax black. Shells black, with a triangular grey spot and two white lines on each; shoulders red-brown.

> *Linn. Syst. Nat.* 2. 639. 18.—*Fn. Sv.* 693.
> LEPTURA, &c. *Geoff. Inf.* 1. 217. 15.
> CERAMBYX albo fasciatus niger, &c. *Degeer. Inf.* 5. 82. 19.
> CERAMBYX quadricolor. *Scop. Ent. Carn.* 177.
> SCARABÆUS, &c. *Raj. Inf.* 83. 26.
> *Schæff. Icon. Tab.* 2. *Fig.* 9.
> CALLIDIUM *mysticum. Fab. Spec. Inf.* 1. 244. 51. 45.

Common

PLATE LXXXIV. 31

Common in the months of *May* and *June*; is usually found in the open path-ways near woods. It appears to be most peculiar to a sandy or light gravel soil.

FIG. III.

LEPTURA AQUATICA.

SPECIFIC CHARACTER.

Green-gold. Antennæ black. A tubercle on each side of the corslet. Shells striated and truncated. Posterior thighs larger with a spine on the interior side.

Linn. Syst. Nat. 2. 637. 1.—*Fn. Sv.* 677.

LEPTURA aquatica spinosa, &c. *Degeer. Inf.* 5. 140. 80. *Tab.* 4. *Fig.* 14. 15.

STENOCORUS, &c. *Geoff. Inf.* 1. 229. 12.

CANTHARIS. *Raj. Inj.* 100. 1.

SCARABÆUS. *Frisch. Inf.* 12. 33. *Tab.* 6. *Fig.* 2.

DONACIA crassipes. *Fab. Spec. Inf.* 1. 245. 52. 1.

This Insect is very common in *England* during the early part of summer; it lives on aquatic vegetables, and runs with much celerity when disturbed. It has also been found among the decayed wood of willow trees.

Fabricius has altered its specific, as well as its generic title; it stands in his System as DONACIA *crassipes.*

FIG.

PLATE LXXXIV.

FIG. IV.

LEPTURA ELONGATA.

SPECIFIC CHARACTER.

Antennæ with black and brown fpots alternately. Head and thorax black. Shells yellow, tipped at the extremity with black; alfo two tranfverfe bands and two fpots of the fame colour. Thighs and part of the legs light brown. Feet black.

Degeer. Inf.

Nearly as rare as the Leptura Arcuata in this country; it is taken in dry fandy places, or among loofe chalk; the foil of *Dartford* and fome other parts of *Kent* is particularly favourable to the increafe of thofe creatures. Met with in the month of *June*.

PLATE

PLATE LXXXV.

PHALÆNA VINULA.

PUSS MOTH.

LEPIDOPTERA.

GENERIC CHARACTER.

Antennæ taper from the bafe. Wings in general contracted when at reft. Fly by night.

SPECIFIC CHARACTER.

Antennæ feathered. Wings grey, ftreaked and waved with dull black; fomewhat diaphanous. Thorax and Abdomen grey fpotted with black.

> *Linn. Syft. Nat.* 2. 815. 29.—*Fn. Sv.* 1112.
> *Geoff. Inf.* 2. 104. 5.
> *Raj. Inf.* 153. 5.
> *Geod. Inf.* 1. *Tab.* 65.
>   2. *Tab.* 37.
> *Merian. Europ. Tab.* 39. *Fig.* 140.
> *Albin. Inf.* 11. *Tab.* 5.
> *Sepp. Inf.* 4. *Tab.* 5.
> *Wilk. pap. Tab.* 13. *Fig.* 1. *e.* 1.
> *Reaum Inf.* 2. *Tab.* 21.
> *Frifch. Inf.* 6. *Tab.* 8.
> *Degeer. Inf.* 1. *Tab.* 23. *Fig.* 12.
> *Roef. Inf.* 1. *phal.* 2. *Tab.* 19.
> *Fab. Spec. Inf.* 2. 178. 52.

The Pufs Moth appears in the winged ftate about the latter end of *May,* or early in *June.*

G   The

PLATE LXXXV.

The Caterpillar, from which it is produced, is of a very extraordinary form, and has rather the appearance of a formidable or venomous creature, than the larva of a Moth : it feeds on Willows and Poplars, and is generally found in great plenty where thofe trees grow, in the month of *July*. The two tails, or crimfon filaments at the extremity of the body, are protruded or concealed within their bafe at the creature's pleafure; when protruded they have a continual writhing or vibratory motion.

It paffes to the Pupa ftate in *Auguft*.

PLATE

86

1

2

3

PLATE LXXXVI.

CARABUS CYANOCEPHALUS.

COLEOPTERA.

GENERIC CHARACTER.

Antennæ taper.   Thorax and fhells margined.   A large appendix at the bafe of the pofterior thighs.   Five joints in each foot.

SPECIFIC CHARACTER.

Thorax and feet orange colour.   Head and fhells blue green.
Linn. Syft. Nat. 2. 671. 21.—*Fn. Sv.* 794.
CARABUS, &c.   *Degeer Inf.* 4. 100. 17. *Tab.* 3. *Fig.* 17.
BUPRESTIS, &c. *Geoff. Inf.* 1. 149. 40.
CANTHARIS, &c. *Raj. Inf.* 89. 1.
Schœff. Icon. Tab. 10. *Fig.* 14.

FIG. I.   The Natural Size.
FIG. II.   The Magnified Appearance of the Upper-fide.
FIG. III.   The Under-fide, Natural Size.

This minute Infect is found in the months of *May* and *June.*

G 2   PLATE

PLATE LXXXVII.

SPHINX FUCIFORMIS.

CLEAR WINGED HUMMING SPHINX.

LEPIDOPTERA.

GENERIC CHARACTER.

Antennæ thickeſt in the middle. Wings, when at reſt, deflexed. Fly flow, morning and evening only.

SPECIFIC CHARACTER.

Antennæ black. Head and Thorax bright yellow; Body rich brown, except the laſt joints, which are yellow; Abdomen bearded with black. Wings tranſparent, with a broad dark brown border; Veins dark.

Linn. Syſt. Nat. 2. 803. 28.—*Fn. Sv.* 1092.

SPHINX, &c. *Geoff. Inſ.* 2. 82.

Roeſ. Inſ. 3. *Tab.* 38.

   4. *Tab.* 34. *Fig.* 1—4.

Bradl. nat. 26. *Fig.* 1. B.

Sulz. Inſ. Tab. 15. *Fig.* 90.

Poda Inſ. Tab. 2. *Fig.* 6.

Schœf. Icon. Tab. 16. *Fig.* 1.

SESIA Fuciformis. *Fab. Sp. Inſ.* 2. 156. 11.

The Caterpillar of this Inſect feeds on the wood of Willows, and is concealed within the ſolid ſubſtance of the trunk, in the ſame

9

manner

manner as the larva of the *Sphinx* Apiformis *, and *Sphinx* Tipuliformis †, are concealed within the wood of the Poplar, and ſtalks of
Currant buſhes.

Fabricius deſcribes the Caterpillar, green with a lateral line of
yellow; ſpine at the end of the body red. *Harris* obſerves, that in
the winged ſtate the fly is found in Gardens, on flowers, in *May*;
Fabricius writes on the Honey-ſuckle, &c.

It is very rare; one ſpecimen has been taken this ſeaſon on *Epping-
Foreſt.*

---

* Plate **XXV.** of this Work.   † Ibid.

PLATE

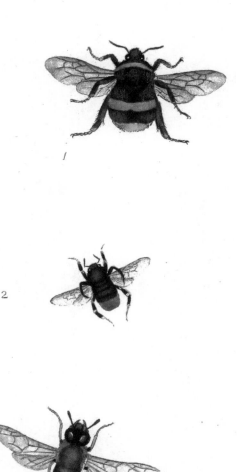

PLATE LXXXVIII

FIG. I.

APIS TERRESTRIS.

HUMBLE BEE.

HYMENOPTERA.

Wings four, generally membraneous. Tail of the females armed with a sting.

GENERIC CHARACTER.

Jaws, with a trunk bent downwards. Antennæ elbowed in the middle; first joint longest. Wings plain. Body hairy. Abdomen connected by a pedicle.

SPECIFIC CHARACTER.

Black, very hairy, with a yellow belt on the Thorax, one also across the Abdomen. Anus white or yellowish.

A. T. hirsuta nigra, thoracis cingulo flavo, ano albo.

Syst. Ent. 379. 5.—*Linn. Syst. Nat.* 2. 959. 41. —*Fn. Sv.* 2709.

Bombylius major niger, linea duplici transversim ducta lutea, alia supra scapulas, alia per medium abdominis, imo abdomine albo. *Raj. Inf.* 247. 5.

Mouff. Inf. 53. t. 2.
Goed. Inf. 2. tab. 46.
Bradl. nat. tab. 26. fig. 1. D.
Reaum. Inf. 6. tab. 3. fig. 1.
Frifch. Inf. 9. tab. 13. fig. 1.

The manners of the common Humble Bee are too well known to require elucidation; its dwelling is formed very deep in the earth; it

H   comes

comes forth when the fun fhines to extract the melliferous moifture of flowers, and is perfectly harmlefs unlefs when irritated. Linnæus defcribes the Anus of the Apis Terreftris white, but I find this is not always conftant; I have feveral fpecimens that agree with the one reprefented in the annexed plate.

I have compared them with the fpecimen in the Linnæan Cabinet; they perfectly agree in every refpect except in the brown or yellow colour of the extreme part of the Abdomen: they are certainly only varieties.

F I G. II.

A P I S　L A P I D A R I A.

R E D - T A I L　B E E.

H Y M E N O P T E R A.

A P I S.

SPECIFIC CHARACTER.

Black, hairy, Anus red-orange colour.
A. L. hirfuta atra, ano fulvo. *Syft. Ent.* 381. 14. *habitat lapidum in acervis.*

Linn. Syft. Nat. 2. 960. 44.—*Fn. Sv.* 1701.—
Geoff. Inf. 2. 417.

Bombylius maximus totus niger, exceptis duobus extremis abdominis annulis rufis. *Raj. Inf.* 246. 1. *Scop. Carn.* 813.
Frifch. Inf. 9. *p.* 25. *Fig.* 2.
Reaum Inf. 6. *t.* 1. *f.* 1. 4.
Schœf. Icon. Tab. 69. *Fig.* 9.

In the Linnæan Cabinet, (now in the poffeffion of Dr. Smith) I find under the name Apis Lapidaria two infects, fo very different in fize, that it certainly will admit fome doubt whether they ought to be

confidered

PLATE LXXXVIII. 43

considered as the same species: Linnæus does indeed, notice this dissimilarity of their size in his description, and says one is three times larger than the other, &c. whence we may conclude that it was after mature deliberation he had ventured to place the smallest as a variety of the other *.—I do not know whether the largest has ever been taken in England; the specimen of it, in the Linnæan Collection, is a Swedish Insect: the smallest (which we have figured) is well known as a native of this country.

It is not found so frequently as most other species of the Apis genus; it lives among heaps of loose stones; its honey is strong.

FIG. III.

TENTHREDO VITELLINÆ.

HYMENOPTERA.

GENERIC CHARACTER.

Abdomen of equal thickness, and closely united to the thorax. Sting serrated, between two valves. Second wings shortest.

SPECIFIC CHARACTER.

Antennæ clavated. Abdomen above black, very hairy, with a lateral line of rufous. Legs yellowish. Thighs behind dentated.

T. V. Antennis clavatis, abdomine supra nigro, lateribus rufis, femoribus posticis dentatis. *Syst. Ent.* 318. 6.

T. V. Antennis clavatis, ore elabiato, abdomine rufo dorso nigro, femoribus posticis dentatis.
Linn. Syst. Nat. 2. 921. 5.—*Fn. Sv.* 1535.
Strœm. Sundm. 171. *Tab.* 10. *Fig.* 11.

---

* " Varietas triplo minor, vix distincta."

Larva

Larva virefcens per aperturam ante anum tanquam e' fiphone aquam exfpuit.   *Fabricius. Spec. Inf.* 1. 407. 7.

The Larva of this fpecies is found on the Alder, Ofier and Willow; it is large, of a green colour, and at firft fight, greatly refembles the Caterpillars of fome Lepidopterous Infects.

When it firft appears from the Chryfalis very little of the black of the Thorax and Abdomen can be feen, thofe parts being at that time thickly cloathed with long brownifh hairs.

PLATE

PLATE LXXXIX.

PAPILIO ANTIOPA.

CAMBERWELL BEAUTY.

LEPIDOPTERA.

GENERIC CHARACTER.

Antennæ clavated. Wings, when at reſt, erect. Fly in the day time.

SPECIFIC CHARACTER.

Wings angulated, rich purple-brown, with a pale yellow external border; and an intermediate dark border, with a row of bluiſh eyes; on the anterior margin of the firſt wings two long yellowiſh ſpots.

——Alis angulatis nigris, limbo albida.
Linn. Syſt. Nat. 2. 776. 165.—Fn. Sv. 1056.
Geoff. Inſ. 2. 35. 1.

Papilio maxima nigra, alis utriſque limbo lato albo cinctis.
Raj. Inſ. 135. 136.
Jonſt. Inſ. t. 9. & 11.
Schœff. Elem. Tab. 94. Fig. 1.
———— Icon. Tab. 70. Fig. 1. & 2.
Sulz. Inſ. 1. Tab. 14. Fig. 85.
Roeſ. Inſ. 1. Pap. 1. Tab. 1.
Eſp. Pap. 1. Tab. 12. Fig. 2.
Seb. Muſ. 4. Tab. 32. Fig. 1, 2.
Bergſtræſs. 2. Tab. 39. Fig. 1. 2. 3. 4.
Wilk. Pap. 58. Tab. 2. a. 10.
Degeer. Inſ. 1. Tab. 21. Fig. 8. 9.

The

The Papilio Antiopa is found in every part of Europe; in Germany particularly it is very common; it is as frequent in America as in Europe, and is efteemed as a rarity only in this country: it is, indeed, fometimes found in abundance with us, but as its appearance is neither annual nor periodical, it is generally valued by Englifh Collectors.

There have been feveral inftances of this Infect being found in different parts of the country in mild feafons, as plenty as the Peacock, or Admirable, Butterflies; in the fummer of 1793 particularly, they were as numerous in fome places as the common garden White Butterfly is ufually near London.

But as a proof that its appearance does not altogether depend on the temperature of the weather, we need only adduce, that not a fingle fpecimen has been taken this feafon, although it has been one of the moft favourable for all kinds of Infects that can be recollected; and many fpecies of Moths and Butterflies, which have not been feen for feveral years before, have been taken at Combe-Wood, Darn-Wood, and fimilar adjacent parts, during fummer, in plenty.

It is from the uncertainty of its appearance that we have fuch different, and, feemingly, irreconcileable accounts of the abundance and fcarcity of this Butterfly; it was certainly well known as a native of this country to former Collectors, yet it received only a few years fince the new name *Grand Surprife*; this name, which was given by Mofes Harris, or by fome of the company of Aurelians, of whofe fociety he was a member, was evidently intended as a fignificant expreffion of their admiration, not of the beauty of the Infect, but of the fingular circumftance of the fpecies remaining fo long in thofe very places where the moft diligent refearches of preceding Collectors had been made in vain; of their unwearied induftry they were well perfuaded, and were therefore unable to account for the appearance of a numerous brood of large Infects, which muft have remained concealed many years, or been lately tranfported to thofe places.

Harris, in his Aurelian, calls it the Camberwell Beauty, though in his lift of Englifh Butterflies Hawk-Moths, and Moths, he ufes the name *Grand Surprife:* we mention this circumftance, as it appears very inconfiftent that the new name he adopts in one work, and the

old

PLATE LXXXIX. 47

old one he fhould have difcarded in the other, are equally and indif-criminately ufed in the feveral editions of both; we ftill find it in the Aurelian, " *Camberwell Beauty*," in the other, " *Grand Surprife*," from which it might be readily inferred, that he meant two diftinct Infects, were it not for the addition of the Linnæan name *Pap. Antiopa*.

In the general defcription of this Infect in the Aurelian, Harris does not fay that it was fcarce at that time (1775), which he certainly would if it had been fo; but Berkenhout, in his outlines of Natural Hiftory, (1789) adds, after its fpecific character, that it is " very rare " in this kingdom." To reconcile thofe accounts, we can only ob-ferve, that no Infect is more uncertain as to the time of its appear-ance; that though found in abundance in one feafon, it may not be feen in the next, or even for feveral fucceffive years; it will then ap-pear in fmall or large quantities, for one, two, or more feafons, and again difappear for many years as before.

The Englifh fpecimens differ from thofe of other countries in the colour of the light exterior border of the wings; in the former, that part is of a very pale yellow brown, inclining to a dirty white; in the latter, it is of a deep yellow, marked and fpotted with brown. *Fabri-cius* notices this difference, and fays they are varieties.

The Caterpillars feed on the Willow, and are generally found on the higheft branches; they caft their fkin early in *July*, and pafs to the Chryfalis, as reprefented in the plate. The underfide of the But-terfly is of a black brown, with irregular dark ftreaks; the yellowifh border is vifible on that fide.

PLATE

PLATE XC.

PHÆLÆNA LŒFLINGIANA.

LEPIDOPTERA.

GENERIC CHARACTER.

Antennæ taper from the bafe. Wings in general contrafted when at reft. Fly by night.

TORTRIX.

SPECIFIC CHARACTER.

Firft wings yellowifh, or buff colour, marked with tranfverfe fhort ftreaks of red, or brick colour, alfo two irregular marks of the fame colour, refembling *XX*, on the anterior margin. Under wings and body lead colour.

P. Alis anticis flavis luteo reticulatis duplici *x x* notatis.
> *Syft. Ent.* 652. 42.
> *Linn. Syft. Nat.* 2. 878. 305.—*Fn. Sv.* 1323.—
> *Clerk. Phal. Tab.* 10. *Fig.* 6.

This little Moth has great affinity with the *Phal. Forfkahliana* of Linnæus, the wings are indeed more angulated, but the form of the *XX* on the upper wings are nearly the fame, and in the general colours both of the upper and under wings they perfectly agree.

Phal. Lœflingiana is found in the greateft abundance on the Oak, in the month of April and May, in the Caterpillar ftate, and in July every Tree that will afford them a moift retreat during the heat of the day, conceals numbers in the winged ftate; morning and evening they are on the wing, they come forth at day break, fport about the bufhes till after fun-rife. and then retire among the thickeft Oak boughs; a little before fun-fet they come forth again, but conceal themfelves as before about twilight.

I

The

The Caterpillars are of a fine green colour, befet with black fpecks, the head is fhining black, a collar of the fame colour paffes round the firft joint, or annulation of the body next the head, but a narrow belt of white paffing between, feparates the black of the head from the fhoulders. It is a brifk creature, and the thread which it fpins is of a very ftrong texture.

It paffes to the Chryfalis ftate in the leaf of the Oak, as fhewn in the plate.

P L A T E

91

PLATE XCI.—XCII.

SPHINX EUPHORBIÆ.

SPOTTED ELEPHANT SPHINX,

LEPIDOPTERA.

GENERIC CHARACTER.

Antennæ thickeft in the middle. Wings, when at reft, deflexed. Fly flow, morning and evening only.

SPECIFIC CHARACTER.

Superior wings light brown, with fpots, and broad ftripes of dark olive. Inferior wings red, marked with black and olive.

Sphinx Euphorbiæ alis integris fafcis, vitta anticis pallida, pofticis rubra. *Syft. Ent.* 541. 17.
> *Linn. Syft. Nat.* 2. 802. 19.—*Fn. Sv.* 1086.—
> *Muf. Lud. Vir.* 356.

Sphinx Euphorbiæ alis integris grifeis, fafciis duabus virefcentibus pofticis rufis bafi ftrigaque nigris, antennis niueis. *Fab. Spec. Inf.* 2. 146. 32.

Sphinx fpirilingius, alis viridi fulvo purpureoque varie fafciatis et maculatis, fubtus purpureis. *Geoff. Inf.* 2. 87. 11.
> *Drury Inf.* 1. *Tab.* 29. *Fig.* 3.
> *Roef. Inf.* 1. *Phal.* 1. *Tab.* 3.
> *Reaum. Inf.* 1. *Tab.* 13. *Fig.* 1. 4. 5. 6.
> *Degeer. Inf.* 1. *Tab.* 8. *Fig.* 6. 11.
> *Schæff. Icon. Tab.* 99. *Fig.* 3. 4.
> *Frifch. Inf.* 2. *Tab.* 11.

SPOTTED ELEPHANT *Harris. Aurel. pl.* 44.

K

The

The Sphinx Euphorbiæ, confidered as a native of this country, is without exception the rareft fpecies of the genus we have : and if we omit the Sp. Porcellus, Lineata, Atropos, with a very few others, we have no indigenous fpecies that can by any means be compared with it as a rare, or, we may add, beautiful Infect.

Drury has given a figure of the Sphinx without its changes among his rare Infects, but as a native of a foreign country : and before the time of *Harris* it was frequently an object of difcuffion among Aure-lians, whether it ever had been taken in *England* ; *Harris* in his work, expreffes himfelf thus, " It has been long in difpute whether " the Spotted Elephant was a native of this ifland ; but it is now paft " a doubt, as I have had the good fortune to find a Caterpillar of this " Moth in marfhy ground at *Barnfcray*, near *Crayford* in *Kent*, about " the middle of *Auguft* * ; it was better than three inches long, of a " dark brown colour ; the horn at the tail part, which was about half " an inch long, appeared long and gloffy. The head was nearly the " fize of a fmall pea, of a lightifh yellow, brown, or tan colour. I " tried various herbs to bring it to feed, but my attempts were fruit- " lefs, and it died for want †. The Chryfalis in the plate was fent " me from *Belleifle* in *France* ; and the Moth was produced from it " about the beginning of *June*."—*Harris*'s *Aurelian*, plate 44.

We are not informed of more than two fimilar circumftances that may place its exiftence in this country beyond difpute ; a damaged fpecimen of the Fly has been taken at *Bath*, and is in our cabinet ; and Mr. *Curtis*, author of the *Flora Londinenfis*, &c. found four of the Caterpillars laft fummer in *Devonfhire*.

In the Caterpillar ftate it frequently changes its fkin, and appears as frequently to alter its appearance ; we cannot elfe account for the diffimilarity that prevails among all the coloured reprefentations of the Infect in that ftate that have come under our infpection ; in Rœfel's Hift. Inf. we find a figure of the Caterpillar apparently in the laft fkin,

---

* 1778.   † It feeds on plants of the *Euphorbia* genus, as its fpecific name indicates.

that

that very nearly correfponds with our fpecimen; but that figured by Harris does not agree with either, in the form or number of the fpots. At an early ftage of its growth the Caterpillar, according to Rœfel, 's bright yellow, with black patches, and minute white fpecks.

The figure in plate XCII. is copied from a moft perfect fpecimen of the Caterpillar, and which is now in our poffeffion; but as we cannot affure our Subfcribers that it was found in *England*, we have been careful to add it in a feparate plate, that fo it may either be included in the volume with the Sphinx and Pupa, or be excluded with propriety.

PLATE

93

1

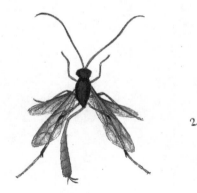

2

PLATE XCIII.

FIG. I.

SPHEX SABULOSA.

HYMENOPTERA.

Wings four, generally membraneous. Tail of the female armed with a fting.

GENERIC CHARACTER.

Jaws, without Tongue. Antennæ of fixteen joints. Wings incumbent, not folded. Sting riged.

SPECIFIC CHARACTER.

Antennæ, Head, Thorax, and Legs black. Abdomen club-fhaped; connected by a flender thread; orange colour; extremity black.

SPHEX SABULOSA. *Syft. Ent.* 346. 1.—*Linn. Syft. Nat.* 2. 941. 1.
—*Fn. Sv.* 1648.

SPHEX, &c. *Degeer Inf.* 2. 2. 148. 4. *tab.* 28. *fig.* 27.
ICHNEUMON, &c. *Geoff. Inf.* 2. 349. 63.
Scop. carn. 770.
Frifch. Inf. 2. *tab.* 1. *fig.* 6. 7.
Sulz. Inf. tab. 19. *fig.* 120.
Schæff. Icon. 83. *fig.* 1.
Fab. Spec. Inf. 2. 442. 112. 1.

Sphex Sabulofa is a very bufy and voracious Infect; it fometimes fubfifts on vegetable matter, frequently on fmall flies; we have never found it common near *London.*

FIG.

F I G. II.

ICHNEUMON CIRCUMFLEXUS.

HYMENOPTERA.

GENERIC CHARACTER.

Jaws, without Tongue. Antennæ of more than thirty joints; long, filiform, vibrating. Sting within a bivalve sheath.

SPECIFIC CHARACTER.

Antennæ, Legs, and Body tawny. Head and Thorax black; as is also the lower part of the second joint of each hind Leg. Body curved.

ICHNEUMON CIRCUMFLEXUS. *Syst. Ent.* 341. 8e.

Linn. Syst. Nat. 2. 938. 59.—*Fn. Sv.* 1631.

Not very common; found in *May* and *June*.

P L A T E

94

1

2

PLATE XCIV

FIG. I.

RHAGIUM BIFASCIATUM.

COLEOPTERA.

RHAGIUM*.

SPECIFIC CHARACTER.

Thorax fpined. Shells olive brown, with three longitudinal ftripes, and two yellow fpots on each.

Fab. Spec. Inf. 1. 230. 4.

Sulz. Hiſt. Inſ. Tab. 5. *Fig.* 8.

Linnæus never defcribed this Infect, or he would have placed it in the *Cerambyx* genus. Fabricius has defcribed it in his Species Infectorum under the fpecific name *Bifafciatum*; but he has feparated it from the Linnæan genus, and given it the new generic title Rhagium : the Cerambyx Inquifitor, C. Curfor and C. Noctis of Linnæus, our prefent fpecies, and R. Ornatum, are the only Infects Fabricius has included in the new genus Rhagium.

The Rhagium Bifafciatum is rare in this country; it is more frequent in France and Germany. It is generally found in putrid flefh.

FIG.

F I G. II.

C E R A M B Y X   M O S C H A T U S.

COLEOPTERA.

GENERIC CHARACTER.

Antennæ articulated, and tapering to the end. Shells long and narrow, four joints in each foot. Thorax with lateral spines or tubercles.

SPECIFIC CHARACTER.

Antennæ length of the body. Shells green, changeable, purple, copper colour, &c. Body dark blue.

CERAMBYX *Moschatus*, Thorace spinoso, elytris obtusis viridibus nitentibus, femoribus muticis antennis mediocribus.
Linn. Syst. Nat. 2. 627. 34.—*Faun. Suec.* 652.

CERAMBYX *odoratus*, &c.   *Degeer. Inf.* 5. 64. 2.

SCARABÆUS.   *Raj.—Frisch.—Lister.*

Few Insects vary more in their colours than the Cerambyx Moschatus; in some specimens the Green colour is very predominant, in others the Copper colour; in some the Purple is the most vivid, and again in others the colours are so blended as to appear altogether of a dull brown. They feed on the soft wood of willow trees; are very plenty in most places in summer, and emit a very powerful musk-like odour.

PLATE

95

PLATE XCV.

PHALÆNA NEUSTRIA.

LACKEY MOTH.

LEPIDOPTERA.

GENERIC CHARACTER.

Antennæ taper from the bafe. Wings, in general, contracted when at reft. Fly by night.

SPECIFIC CHARACTER.

Antennæ feathered. Head, Thorax, Body, and Wings light brown; a dark broad wave acrofs the middle of the upper Wings.

P. Neuftria. B. alis reverfis grifeis, ftrigis duabus ferrugineis, fubtus unica. *Syft. Ent.* 567. 42.—*Linn. Syft. Nat.* 2. 818. 35.

Phalæna pectinicornis elinguis, alis deflexis pallidis, fafcia alarum tranfverfali faturatiore. *Geoff. Inf.* 2. 114. 16.

Phalæna media tota cinerea. *Raj. Inf.* 214. 8.
　　　　　　　Reaum. Inf. 2. *Tab.* 4. *Fig.* 1.—11.
　　　　　　　Goed. Inf. 1. 57. *Tab.* 10.
　　　　　　　Harris's Aurel. pl. 17.
　　　　　　　Wilk. Pap. 21. *Tab.* 3. *a* 10.
　　　　　　　Alb. Inf. 19. *Fig.* 27.
　　　　　　　Frifch. Inf. 1. *Tab.* 2.
　　　　　　　Roef. Inf. 1. *Phal.* 2. *Tab.* 6.
　　　　　　　Fab. Spec. Inf. 2. 180. 58.

The

The Caterpillar of the Ph. Neuſtria are found in June, either on the white-thorn, black-thorn, or briar; ſometimes on fruit trees: they paſs to the Chryſalis ſtate in July, and the Moths appear in Auguſt.

The female depoſits her eggs with ſuch particular care and regularity, that a cluſter of them forms one of the moſt pleaſing objects for microſcopical inveſtigation; they are cruſtaceous, of a light grey or bluiſh colour, elegantly marked at the broadeſt end; they are diſpoſed with the greateſt ſymmetry around the ſmall branches of the thorn, and are ſo cemented together that they cannot readily be ſeparated.—The appearance of a cluſter is repreſented in our plate.

The eggs are laid in autumn, though they are not hatched till the enſuing ſpring. When the young Caterpillars burſt forth, they form into ſocieties, ſometimes of thirty or forty individuals, ſometimes of a much greater number; they immediately commence the formation of a ſpacious web, and if the weather be fine in two or three days, their work is completed; as however they encreaſe in bulk, it is neceſſary to enlarge their dwelling, and this they manage either by adding new external coverings, or encreaſing and extending the windings within. They ſeldom paſs to the Pupa form in thoſe neſts, but ſeparate in ſearch of a more convenient place for that purpoſe when they have attained their full ſize.

The Caterpillar, when preparing for its next ſtate, weaves a large ſilky caſe; within which it forms another ſomewhat ſmaller; and thus enveloped by its double cone, it changes to the Pupa form. The Pupa is black, and may be juſt diſcerned through the two caſes, as repreſented in our plate.

The figure of the perfect Inſect is copied from a female ſpecimen; the male is rather darker, and has the Antennæ more feathered.

PLATE

96

1

4

3

2

PLATE XCVI.

FIG. I.

CHRYSOMELA POLYGONI.

COLEOPTERA.

GENERIC CHARACTER.

Antennæ knotted, enlarging towards the ends. Corselet margined.

SPECIFIC CHARACTER.

Head, Shells, and underside blue green. Thorax and Thighs
orange colour. Globules of the Antennæ of equal size.

C. Polygoni. Ouata cærulea, thorace femoribus anoque rufis.
Syst. Ent. 100. 32. — *Linn. Syst. Nat.* 2, 589. 24.—
Fn. Sv. 520,

Chrysomela, &c. Geoff. Inf. 1. 283. 4.
Chrysomela, &c. Degeer. Inf. 5. 322. 26.
Reaum. Inf. 3. *Tab.* 17. *Fig.* 14. 15.
Schæff. Icon. Tab. 51. *Fig.* 5.
Tab. 161. *Fig.* 4.
Tab. 173. *Fig.* 4.

This pretty, though common Insect, is generally found on those
plants which grow on the banks of ditches in the months of May or
June.

FIG.

FIG. II.

CANTHARIS ÆNEA.

GENERIC CHARACTER.

Antennæ taper. Thorax margined. Shells flexile. Sides of the Abdomen papillous, and folded. In each Foot five joints.

SPECIFIC CHARACTER.

Bright green. Shells red on the external fides; a fmall red fpot on each fide of the Corfelet.

CANTHARIS *Ænea* thorace marginato, corpore viridi æneo elytris extrorfum undique rubris. *Linn. Syft. Nat.* 2. 648. 7.—
Fn. Sv. 708.

Cicindela viridi ænea, elytris extrorfum rubris.
Geoff. Inf. 1. 174. 7.

Thelephorus æneus, &c. *Degeer. Inf.* 4. 73. 6. *Tab.* 2. *Fig.* 16. 17.

Scarabæus, &c. *Raj. Inf.* 77. 12.
Schæff. monogr. 1754. *Tab.* 2. *Fig.* 10. 11.
Icon. Tab. 18. *Fig.* 12. 13.

Very plenty on flowers; often on thiftles in May.

FIG.

PLATE XCVI.   65

FIG. III.

STAPHYLINUS MAXILLOSUS.

COLEOPTERA.

GENERIC CHARACTER.

Antennæ globular. In each Foot five joints. Shells curtailed. Wings covered. Tail defencelefs, with two veficles.

SPECIFIC CHARACTER.

Black. Antennæ of eleven globules. Jaws as long as the Head. Shells grey, cover one third of the Abdomen. Length one inch.

Sp. Maxillofus. Pubefcens niger, fafciis cinereis. *Syft. Ent.* 265. 3. *Linn. Syft. Nat.* 2. 683. 3.—*Fn. Sv.* 841.

Staphylinus, &c. *Geoff. Inf.* 1. 360. 1. *Tab.* 7. *Fig.* 1.

Staphylinus balteatus, &c. *Degeer. Inf.* 4. 18. 4. *Tab.* 1. *Fig.* 7. 8.

Scarabæus. Lift. Logu. 391.
   Jonft. Inf. Tab. 17. *Fig.* 1. 2. 3.
   Bocc. Muf. 2. *Tab.* 31. *Fig.* AA.
   Schæff. Icon. Tab. 20. *Fig.* 1.
Staphylinus olens, &c. *Müll. Faun. Fridrickfd.* 23. 228.
     Zool. Dan. 97. 1090.

Found chiefly in fandy places ; may be often obferved flying againft dry banks when the fun fhines ; makes a buzzing noife ; feeds on decayed vegetables, but more efpecially on the flefh of dead animals. Met with in May, June, and July.

FIG.

FIG. IV.

ELATER SPUTATOR.

COLEOPTERA.

GENERIC CHARACTER.

Antennæ taper, lodged in a groove under the Head and Thorax. Under fide of the Thorax terminates in a point lodged in a cavity of the Abdomen. Spring to a confiderable height when laid on their backs.

SPECIFIC CHARACTER.

Thorax black.   Shells brown.   Body black.

Linn. Syft. Nat. ed. 12. 182. 15.
Faun. Suec. 583.

We have feveral fpecies of this genus that fo nearly refemble each other, as fcarcely to be diftinguifhed on the moft accurate inveftigation from the E. Sputator. They are found in great abundance in fummer.

PLATE

97

PLATE XCVII.

PHALÆNA LUCIDATA.

DARTFORD EMERALD MOTH.

LEPIDOPTERA.

GENERIC CHARACTER.

Antennæ taper from the bafe. Wings in general contracted when at reft. Fly by night.

SPECIFIC CHARACTER.

Fine lucid green, two white waves acrofs the upper, and one acrofs the under wings.

This fpecies we have ever found peculiar to the woods about two or three miles beyond Dartford (Kent), particularly on the fkirts of Darnwood, and near the banks of the river Thames at Queenhithe; it has probably never been taken elfewhere, or the name Dartford Emerald would not have been fo generally adopted by Collectors.

It is not very frequent even in thofe local fituations, nor can we learn that its larva and pupa ftate has been afcertained before; the fpecies has neither been defcribed by *Linneus* nor *Fabricius*; *Harris* does not mention it in his catalogue of Englifh Moths, nor has a figure of it been given in any preceding publication that have come under our infpection.

The fpecific name is intended to exprefs the lucid or tranfparent appearance of the Infect.

M   I am

I am not certain whether in the larva ftate it feeds on the Convolvulus, although I found it on a plant of that kind; as its climbing ftalks and tendrils were fo intricated with branches of white-thorn, oak, and broom, as to preclude any accurate determination.

I kept them in a gauze cage for the fpace of a fortnight, and fupplied them with freſh portions of the different plants every day, but could never obferve them take the leaft fubfiftence during the whole time; they affixed their tails and hinder legs in the meſhes of the gauze when I firſt removed them into the cage, and never ſhewed the leaft figns of life after; as they held firmly by the gauze, in the pofitions reprefented in our plate, I was very much difappointed to find on attempting to remove them, that two were dead; May 23d I obferved that which was alive threw out a very delicate white thread, as if about to fpin a cone; the body gradually ſhrivelled at the upper part, while the lower became proportionably thicker; two days after it fell to the bottom of the cage and became a pupa, at firſt of a whitiſh, and after of a fine green colour, marked at the narrow end with ſhort black ſtreaks. June 13th the Moth came forth.

At Fig. I. is ſhewn the head of the Caterpillar magnified; it is grey, with the jaws black, and is concealed beneath two horns or projections of the fame green colour as the back.

P L A T E

98

PLATE XCVIII.

CIMEX LURIDUS.

HEMIPTERA.

Shells or upper wings, femi-cruftaceous, not divided by a ftraight future, but incumbent on each other. Back curved downwards.

GENERIC CHARACTER.

Antennæ longer than the thorax. Thorax margined, in each foot three joints.

SPECIFIC CHARACTER.

Thorax fpined, brown, tinged with green. Shells brown, with a dark fpot on the center of each.

Cimex *Luridus.* Thorace obtufe fpinofo fubvirefcente, elytris grifeis, macula fufca, clypeo emarginato.

Syft. Ent. 701. 25.
Fab. Spec. Inf. 2. 345. 38.

Fabricius is the only writer who has defcribed this beautiful Infect; the defcription in the Species Infectorum is taken from a fpecimen in the collection of *Sir J. Banks, Bart.* A very minute Latin account is alfo given in a Mantiffa of Entomology lately publifhed by the fame author, but in which he does not even mention the larva or pupa ftate, though their characters differ fo effentially from the perfect Infect; we fufpect in the two firft ftates the Infect has hitherto re-

M 2   mained

mained unknown, as in the perfect state it is very rarely met with., We have never seen a figure of either in any former publication.

June 10th, 1794.—I found one specimen in the larva state at Coombe-wood, Surrey; it was lurking beneath a branch of hazel, among some small Caterpillars that had formed a slight web on the leaves; as it was only served with vegetable food when confined in the breeding cage, it died in a few days.

June 26th, 1794.—I shook another specimen from the upper branches of a tall oak in Darn-wood, Dartford. At first it refused to eat, but shortly after I observed it suspended acrofs a leaf, with its head downward, and its rostrum extended and transfixed through the head of a small Caterpillar which had unfortunately strayed into the box. I fed it after with dead worms, house flies, &c. from which it extracted nutritive moisture, and encreafed confiderably in bulk.—June 29th it cast its exuviæ—July the 2d. it cast another, when the perfect Insect came forth: the larva can scarcely be distinguished from the pupa state.

Fig. I. the natural size of the larva, with its manner of feeding.—underside.

Fig. II. magnified appearance of the upperside of ditto.—The perfect state shewn above.

P L A T E

99

2

PLATE XCIX.

CHRYSOMELA BILITURALA.

COLEOPTERA.

Wings two, covered by two ſhells, divided by a longitudinal future.

GENERIC CHARACTER.

Antennæ knotted, enlarging towards the end. Coᵣſelet margined.

SPECIFIC CHARACTER.

Antennæ near the length of the body, black. Head, thorax, and underſide, black. Shells red, inclining to yellow brown, with a broad longitudinal black ſtripe extending from the baſe, nearly to the extremity of each.

This Inſect is deſcribed in the manuſcripts of T. MARSHAM, ESQ. S. L. S. who favoured me with the ſpecimen from which the figure in the annexed plate is copied; it does not appear to have been either figured or deſcribed in any preceding Natural Hiſtory, and may therefore be eſteemed as a rare Inſect. The ſpecific name *biliturala* is adopted from that Gentleman's manuſcripts by permiſſion.

Is found on Hornbeam in May.

100

PLATE C.

PHALÆNA CŒRULEOCEPHALA.

FIGURE OF EIGHT MOTH.

LEPIDOPTERA.

GENERIC CHARACTER.

Antennæ taper from the base. Wings in general deflexed when at rest. Fly by night.

SPECIFIC CHARACTER.

Antennæ feathered. Superior wings brown, marbled with blueish green; the resemblance of a double figure of eight on each. Inferior wings lighter with a brownish scallopped margin.

PHALÆNA CŒRULEOCEPHALA elinguis cristata, alis deflexis griseis, stigmatibus albidis coadunatis.—*Linn. Syst. Nat.* 2. 826. 59.—*Fn. Sv.* 1117.

PHALÆNA pectinicornis elinguis, alis deflexis fuscis, macula duplici albo flavescente, geminata. *Geoff. Inf.* 2. 122. 27.

> *Raj. Inf.* 163. 17.
> *Goed. Inf.* 1. *tab.* 61.
> *Reaum. Inf.* 1. *tab.* 18. *fig.* 6. 9.
> *Roef. Inf.* 1. *phal.* 2. *tab.* 16.
> *Frisch. Inf.* 10. *tab.* 3. *fig.* 4.
> *Merian. Europ. tab.* 9.
> *Albin. Inf. tab.* 13. *fig.* 17.
> *Wilks Pap.* 6. *tab.* 1. *a* 12.
> *Haris. Aurel. pl.* 30. *a. b. c. d.*
> *Fab. Spec. Inf.* 2. 184. 72.

N   The

The Caterpillars of this species are found in their laſt ſkin about the latter end of May, or early in June; they change into chryſalis a few days after. The Moth is produced in Auguſt.

In the Caterpillar ſtate they are met with in great plenty, either on the crab tree, black thorn, or white thorn; but are not ſo abundant in the laſt ſtate, as many periſh when in chryſalis.

They change into chryſalis within a hard caſe, which they faſten to the ſmall ſtems of trees.

PLATE

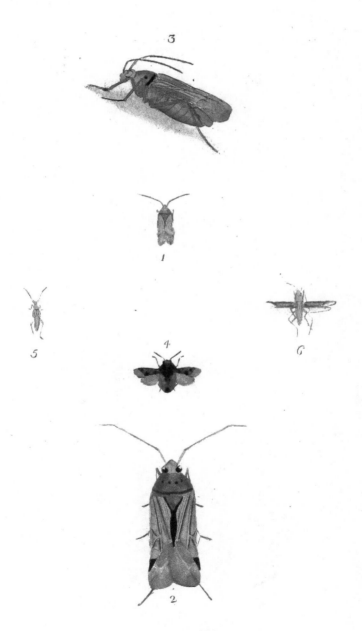

PLATE CI.

CIMEX.

HEMIPTERA.

Shells, or upper wings, femi-cruftaceous, not divided by a ftraight future, but incumbent on each other. Beak curved downward.

GENERIC CHARACTER.

Antennæ longer than the thorax. Thorax margined. In each foot three joints.

FIG. I. II. III.

CIMEX QUADRIPUNCTATUS.

SPECIFIC CHARACTER.

Antennæ yellow. Eyes black. Head and thorax yellowifh orange colour; four diftinct black fpots, and a tranfverfe band of the fame on the latter. Wings yellow, with an orange fhade, and ftreaked with black. Legs and body bright orange.

This very rare and non-defcript fpecies is diftinct from the *Cimex ftriatus,* with which it has been fuppofed to have fome affinity; it is fmaller, the head, thorax, and body are very different, though in the colours of the wings they nearly correfpond.—The four black fpots on the thorax furnifh our fpecific diftinction.

Fig. I. natural fize. Fig. II. and Fig. III. the Infect magnified.

FIG.

PLATE CI.

FIG. IV.

CIMEX FESTIVUS.

SPECIFIC CHARACTER:

Head, thorax, body and fhells red, with black fpots; fix black fpots on the thorax. Inferior wings pale brown.

C. Festivus. Ovatus nigro rubroque varius, thorace punctis fex nigris, alis fufcis, margine albido. *Fabric. Syft. Ent.* 714. 87.
Linn. Syft. Nat. 2. 723. 57.
Cimex dominulus. *Scop. Carn.* 362.
Fuefly Inf. Helv. 26. 490.
Die Staatfwanze. *Panzer Faun. Inf. Germ.* 6. 19.

The Cimex feftivus is very rarely taken in this country. Our fpe‑cimen was found on a ftrawberry bed in june 1794.

FIG. V. VI.

CIMEX PALLESCENS.

SPECIFIC CHARACTER.

Linear. Upper and under wings very pale brownifh colour. Thorax and body pale yellow with two faint crimfon longitudinal ftreaks from the antennæ to the extreme part of the body.

This little Infect is defcribed in the manufcripts of T. Marfham, Efq. s. l. s. under the fpecific name C. Pallefcens; it is by no means uncommon though it has never appeared in any former publication.

-394-

In

PLATE CI. 79

In the larva and pupa ftate it is a very beautiful creature, as the colours are much brighter than in the perfect Infect; they are generally found in April or May, among the grafs and young plants that grow under hedges; in June or July they are taken in the winged ftate.—Fig. V. the pupa ftate, and Fig. VI. the perfect Infect; both of the natural fize: in the annexed plate we have given the magnified appearance of the former.

PLATE

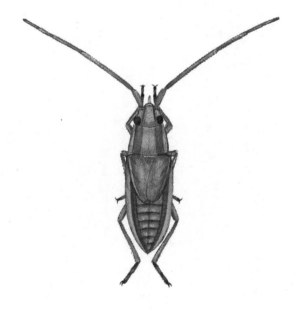

PLATE CII.

THE

LARVA

OF

CIMEX PALLESCENS

MAGNIFIED.

PLATE CIII.

PHALÆNA QUERCUS.

LARGE EGGER MOTH.

LEPIDOPTERA.

GENERIC CHARACTER.

Antennæ taper from the bafe. Wings, in general, contracted, when at reft. Fly by night.

SPECIFIC CHARACTER.

Antennæ of the Male feathered. Wings dark brown, with a bright yellow bar acrofs each, and a ftrong white fpot on the center of each fuperior wing.—Female marked like the Male, but of a paler colour.

PHALÆNA QUERCUS. *Linn. Syft. Nat.* 2. 814. 25.—*Fn. Sv.* 1106.

PHALÆNA maxima fulva, alarum exteriorum fuperioritate intenfius colorata, cum macula in media alba, inferiore dilutiore. *Raj. Inf.* 142. 2.

Merian. Europ. 1. *tab.* 10.
Harris. Aurel. pl. 29. *a. b. c. d. e. f.*
Albin. Inf. tab. 18. *fig.* 25.
Reaum. Inf. 1. *tab.* 35.
Ammiral. Inf. tab. 31.
Roef. Inf. 1. *phal.* 2. *tab.* 35.
Petiv. Gazoph. tab. 45. *fig.* 5.
Goed. Inf. 1. 51. *tab.* 7.

O

The

The Caterpillars of this Moth feed on the White and Black Thorn, together with several herbaceous plants; it has been observed to thrive better in the breeding cage when regularly supplied with fresh grass, to keep the former in a proper state of moisture.

The Female deposits her eggs in June or July, the Caterpillars are hatched in Autumn, and remain in that state during the Winter; about the middle of May it spins a large brown case, within which it passes to the Pupa state; the Moths appear in June.

In the Caterpillar state it is scarcely possible to distinguish the Male from the Female, except that the former is smaller than the latter; but in the last state their colours are entirely different, the Female being of a pale yellowish teint, inclining to fox colour, the Male is of a rich brown.

The Eggs are very curious, they resemble in shape those of a Hen, but are neatly mottled with dark brown.

The Caterpillars cast their skins several times, and always thereby assume a new appearance, though the general colours and character of the species may be traced through every stage. Our figure is copied from a very large and fine coloured specimen of the Female, that was met with at *Darent-Wood, Dartford.*

Explanation of the Figure shewn in Plate 103.

The Eggs of the natural size.

The Case which encloses the Pupa; the former is torn open to expose the latter within.

PLATE

1

2

PLATE CIV.

PHALÆNA QUERCUS.

IN THE

WINGED STATE.

FIG. I. The Male.

FIG. II. The Female.

PLATE

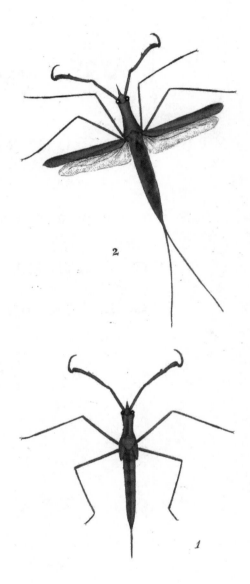

PLATE CV.

NEPA LINEARIS.

LINEAR WATER SCORPION.

HEMIPTERA.

GENERIC CHARACTER.

Antennæ, or Fore-legs cheliform. Wings croffed and complicated.

SPECIFIC CHARACTER.

Brown, cylindrical. Head fmall. Thorax long. Legs four. Abdomen red, with two long tails.

NEPA LINEARIS, manibus fpina laterali pollicatis. *Linn. Syft.*
Nat. 2. 714. 7. *Fn. Sv.* 908.

NEPA LINEARIS corpore anguftiffimo elongato, thorace longo,
tibiis anticis in medio fpina laterali. *Degeer. Inf.* 3.
369. 2. *tab.* 19 *fig.* 1. 2.

Locufta aquatica, *Mouffeti. Raj. Inf.* 59.
     Fuef. Inf. Helv. 25. 473.
     Gronov. Zooph. 683.
     Schœff. Icon. tab. 5. *fig.* 56.
     Swammerdam Bibl. Nat. 1. 233; *tab.* 3.
     fig. 9.
     Roef. Inf. 3. 141. *tab.* 23.

This fingular fpecies is by no means fo common as the *Nepa*
Cinerea, already figured in this Work. One fpecimen was taken at
Ilford, in *Effex,* laft September; and *Thomas Walford,* Efq; met

3
                    with

with another in a bog near *Clare Priory*, *Suffolk* : the latter is preferved in the Mufeum of Mr. *Parkinfon*.

In the Larva and Pupa ftate it is very rarely met with, as it lives
in deep ftagnant water ; the figure of the latter, which we have given
at Fig. I. is copied from the only Englifh Specimen of the Infect we
have ever feen in that ftate; it was taken out of a Pool, near *Epping*,
in the month of June, 1790.

Fig. 2, the perfect Infect.

PLATE

106

PLATE CVI.

FIG. I.
FIG. V?

PHALÆNA EMARGANA.

Notch Wing.

Lepidoptera.

GENERIC CHARACTER.

Antennæ taper from the bafe. Wings in general contracted when at reft. Fly by night.

Tortrix, *Linnæus.*
Pyralis, *Fabricius.*

SPECIFIC CHARACTER.

Firft wings brown, with reticulated dark lines; the anterior margin deeply excavated in the form of a curve. Pofterior wings light brown.

P. *Emargana.* Alis fubcaudatis flavis fufco reticulatis fafciaque lata fufca, margine craffiori late emarginato. *Syft. Ent. Fab.* 651. 37.

The excavations of the fuperior wings of this Infect contribute such an air of novelty to its general appearance, that it might rather be confidered as the effect of chance or defign, on a fingle Infect, if we did not obferve that character prevail through every fpecimen; we find two kinds of them in feveral cabinets in London, and we are in poffeffion of a third that differs from either.

P   By

By moſt practical Entomologiſts they have been conſidered as diſtinct ſpecies, and they may be ſuch; but as we are unwilling to create
confuſion by extending the number of ſpecies, we prefer admitting
them as varieties under the Fabrician name Emargana.—We are more
readily inclined to adopt this meaſure, as we have always found them
at the ſame time of the year, in the ſame parts of the woods, and
generally ſporting together, which is not commonly obſerved of Inſects
that are not either varieties or differ only in ſex.

They are rarely met with; our ſpecimens were taken in June at
Dartford.—They have been taken together at Caen-Wood, Hampſtead.

They have not been deſcribed by Linnæus; but were known among
Collectors by the (now obſolete) name Excavana.

At Fig. 5. is ſhewn one of the varieties; the third is much yellower
but marked with ſimilar reticulated ſtrokes, and exactly correſponds
in ſize and form with this figure.

FIG. II.

PHALÆNA ZOËGANA.

LEPIDOPTERA.

TORTRIX.

SPECIFIC CHARACTER.

Firſt wings yellow, with a brown ſpot on the middle; exterior of
each dark brown, with a large ſplaſh of yellow in the center. Second
wings dark brown.

PHAL. *Zoëgana* alis flavis puncto medio furrugineo, poſtice
ferrugineis macula flava. *Linn. Syſt. Nat.*
2. 876. 289.

β. PHAL.

PLATE CVI.

93

β. PHAL. *hamata* alis fuperioribus flavis puncto lituraque poftica hamata ferrugineis. *Linn. Syft. Nat.* 2. 876. 290.

Fn. Sv. 1309.

Clerk. Phal. tab. 4. *fig.* 4.

——————— *tab.* 4. *fig.* 5. 6.

Fabri. Spec. Inf. 2. 280. 25.

Taken on Epping Foreft in June.—We have rarely met with this Infect.

FIG. III.

PHALÆNA QUERCANA.

LEPIDOPTERA.

TORTRIX.

SPECIFIC CHARACTER.

Antennæ very long. Firft wings pale pink, margined with yellow: yellow fpots on the center. Inferior wings pale; underfide tinctured with pink.

PHAL. *Quercana* alis anticis flavis, maculis daubus coftalibus fulphureis. *Fab. Syft. Ent.* 652. 39.

PHALÆNA *fagana Wien.* Vers. 28. 7. *tab.* 1. *a. b.*

——————— *tab.* 1. *b. b.*

The low oaks, and particularly fuch as are encircled with ivy, generally afford a fhelter to numbers of this pretty Infect during the heat of the day; they are feldom found in the thickeft of the wood, they feem to prefer the thick hedges by the road fides.

Is found in the months of May, June, and July.

FIG.

FIG. IV.

PHALÆNA PANZERELLA.

LEPIDOPTERA.

TINEA.

SPECIFIC CHARACTER.

Long, narrow. Anterior wings pale clay colour, with a dark
ſtreak down the middle, and a few minute ſpots of the ſame colour
near the apex. Poſterior wings almoſt tranſparent, bluiſh, fringe
very deep, of a clay colour.

This elegant Inſect was found the latter end of autumn, 1794,
among ſome high graſs and water plants in the vicinity of Hampſtead,
and is now in the poſſeſſion of the author.

It has certainly never been deſcribed or figured before ; nor is it in
the cabinet of any Entomologiſt within the circle of our friends; if
we except a very diſtinct variety which is in the cabinet of Mr. *Honey,*
Union-Street, Borough.

We have named it Panzerella in honour of the German Entomologiſt DR. GEORGE WOLFFGANG FRANZ PANZER, Author of Faunæ
Inſectorum Germanicæ initia, &c.

     P L A T E

107

1

P L A T E CVII.

C U R C U L I O   A R G E N T A T U S.

COLEOPTERA.

Wings two, covered by two fhells, divided by a longitudinal future.

GENERIC CHARACTER.

Antennæ clavated, elbowed in the middle, and fixed in the fnout, which is prominent and horny.   Joints in each foot four.
* * Snout fhort.   Thighs dentated.

SPECIFIC CHARACTER.

Covered with fine green bronze fcales.   Antennæ and legs brown.

C. breviroftris femoribus dentatus ; corpore viridi argenteo. *Syft. Ent.* 155. 148. *Linn. Syft. Nat.* 2. 615. 75.

Curculio fquamofus, viridi auratus.   *Geoff. Inf.* 1. 293. 38.

Curculio *Urticæ,* &c.   *Degeer. Inf.* 5. 219. 12.
Sulz. Hift. Inf. tab. 4. *fig* 9.
Fab. Spec. Inf. 1. 198. 218.

This elegant little Infect is very common during the fummer in almoft every fituation.   It generally appears in abundance in May and June.

At Fig. I. is fhewn the natural fize.
Fig. II. the magnified appearance.

PLATE

108

2

1

PLATE CVIII.

FIG. I.

APIS LAPIDARIA.

LARGE RED-TAIL BEE.

HYMENOPTERA.

Wings four, generally membraneous. Tail of the Female armed with a sting.

GENERIC CHARACTER.

Jaws, with a trunk bent downwards. Antennæ elbowed in the middle, first joint longest. Wings plain. Body hairy. Abdomen connected by a pedicle.

SPECIFIC CHARACTER.

Entirely black except the tail, which is red.

Linn. Syst. Nat. 2. 960. 44.
Fn. Sv. 1701.
Geoff. Inf. 2. 417.
Fabri. Spec. Inf. 1. 477. 17.

In Plate LXXXVIII. of this work I gave a figure of the Small Apis Lapidaria, Red-tail Bee, which is well known as a native of this country; but declined including a figure of the largest kind, until I could affirm on credible authority it had been taken in England also.

I have lately had the good fortune to be satisfied in this particular; LORD WILLIAM SEYMOUR favoured me with the specimen from

3 which

which the annexed figure is copied; his Lordfhip told me he met with it in Wiltfhire laft fummer, with feveral other rare Infects, which will appear fhortly in this work.

F I G. II.

A P I S   A C E R V O R U M.

BLACK BEE.

H Y M E N O P T E R A.

APIS.

SPECIFIC   CHARACTER.

Entirely Black.   Hairy.

APIS *Acervorum* hirfuta atra.   *Linn. Syft. Nat.* 2. 261. 50.
Fn. Sv. 1717.
Schæff. Icon. tab. 78. *fig.* 5.

This fpecies lives in the earth, it is not often met with near London, We received it through the fame channel as the former.

LINNÆAN INDEX

TO

VOL. III.

COLEOPTERA.

Q

Noto-

INDEX.

LEPIDOPTERA.

* The Star * diftinguifhes thofe which have not been named before.

NEUROP-

I N D E X.

NEUROPTERA.

HYMENOPTERA.

ALPHABETICAL INDEX

TO

VOL. III.

INDEX.

9

ERRATA

ERRATA to Vol. III.

Figures on the Plate annexed to Page 19—" for Plate LXXVIII, read Plate LXXIX."

Plate XCVII, page 67, line 11, for Darnwood, read Darentwood.

———————————— line 13, for Queenhithe, read Greenhithe.

Plate XCIX, for C. Biliturala, read Biliturata.

THE

NATURAL HISTORY

OF

BRITISH INSECTS;

EXPLAINING THEM

IN THEIR SEVERAL STATES,

WITH THE PERIODS OF THEIR TRANSFORMAT ONS,
THEIR FOOD, OECONOMY, &c.

TOGETHER WITH THE

HISTORY OF SUCH MINUTE INSECTS

AS REQUIRE INVESTIGATION BY THE MICROSCOPE.

THE WHOLE ILLUSTRATED BY

COLOURED FIGURES,

DESIGNED AND EXECUTED FROM LIVING SPECIMENS.

By E. DONOVAN.

VOL. IV.

LONDON:

PRINTED FOR THE AUTHOR,

And for F. and C. RIVINGTON, Nº 62, ST. PAUL'S CHURCH-YARD,

MDCCXCV.

109

THE
NATURAL HISTORY
OF
BRITISH INSECTS,

PLATE CIX.

PAPILIO PODALIRIUS,

SCARCE SWALLOWTAIL.

BUTTERFLY.

LEPIDOPTERA.

GENERIC CHARACTER.

Antennæ clavated.   Wings when at reſt erect.   Fly by day.

SPECIFIC CHARACTER.

Above pale yellow, beneath paler. On the firſt wings (upper ſide) ſix pale black ſtripes and a black margin. On the ſecond wings, an oblique black ſtripe, and a black border with five ſemilunar blue ſpots, two long tails.   Stripes more numerous on the under ſide.

PAPILIO *Podalirius*, alis caudatis ſubconcoloribus flaves centibus, faſciis fuſcis geminatis, poſticis ſubtus linea ſanguinea.

Syſt. Ent. 451. 38.
Linn. Syſt. Nat. 2. 751. 36.
Muſ. Lud. Vir. 208.

B 2   *Papilio*

4

Papilio alis pallide flavis, rivulis tranfverfis nigris fecundariis angulo fubulato maculaque crocea. *Geoff. Inf. 2. 56. 24.*

> *Papilio Sinon. Pod. Inf. 62. tab. 2. fig. 1.*
> *Cram. Inf. 13. tab. 152. tab. 2. fig. 1.*
> *Merian. Europ. 163. tab. 44.*
> *Roef. Inf. 1. pap. 2. tab. 2.*
> *Reaum. Inf. 1. tab. 11. fig. 3. 4.*
> *Jonft. Inf. tab. 5. fig. 5.*
> *Efp. pap. 1. tab. 1. fig. 2.*
> *Schæff. elem. tab. 94. fig. 4.*
> ——— *Icon. tab. 45. fig. 3. 4.*
> *Raj. Inf. 111. 3.*
> *Fab. Spec. Inf. 2. 15. 58.*

Fabricius * and fome other entomological writers have very minutely defcribed the Larva and Pupa ftate of this rare butterfly; the Larva feed on the leaves of the turnip, cabbage, and other plants of the fame genus; it is of a yellow colour, with fpots of brown, head pale green. The Pupa is yellow, fpotted with brown alfo, and has two teeth, or fharp points in the fore-part.

We have received the Butterfly from North America, as well as from Germany; it appears to be a native of moft parts of the Euro-pean Continent, though perhaps not frequently found.—*Berkenhout* is the only writer who has defcribed it as an Englifh fpecies †; he fays it is rare (in this country,) found in woods. In the perfeft ftate, vifits flowers in the day time.

---

* Habitat in Europæ Braflicæ.
Larva folitaria, flavefcens, fufco punctata, capite pallide virefcente.
Puppa flavefcens, fufco punctata, antice bidentata. *Fab. Spec. Inf. &c.*

† Synopfis of the Natural Hiftory of Great Britain and Ireland.

P L A T E

110

PLATE CX.

PHALÆNA PENTADACTYLA.

WHITE FEATHERED MOTH.

LEPIDOPTERA.

GENERIC CHARACTER.

Antennæ taper from the base. Wings in general contracted when at rest. Fly by night.

* 7 * ALUCITÆ.

SPECIFIC CHARACTER.

Every part snow white, except the eyes, which are black, anterior wings bifid, posterior tripartite.

PHALÆNA PENTADACTYLA *Alucita* alis patentibus fissis quinque partitis niveis, digito quinto distincto. *Lin. Syst. Nat.* 2. 542. 304. *edit.* 10.

P. Pentadactylus, alis niveis, anticis bifidis, posticis tripartitis.

Syst. Ent. 672. 6—*Fab. Spec. Inf.*
Geoff. Inf. 2. 91. 1.
Reaum. Inf. 1. *tab.* 20. *fig.* 1. 2.
Roef. Inf. 1. *phal.* 4. *tab.* 5.
Ammir. Inf. tab. 23.
Sulz. Inf. tab. 16. *fig.* 10.
Petiv. Gazoph. tab. 67. *fig.* 6.

The Caterpillar of this singular Insect is very common in May; it is of a green colour, with a white stripe down the back, and one on each side; it casts its skin several times.

We have observed some Caterpillars which were quite smooth, after casting their skin become rough or covered with hairs; and others which

6

were

were white become black by the fame procefs ; in this caterpillar we
have obferved a fimilar change : a fpecimen which was of a plain green
as before defcribed, became fuddenly fpotted with black as fhewn in
our plate, that fkin being caft off it affumed its former appearance
and became a pupa.

It feeds on grafs, nettles, &c. near the fides of ditches, and is found
fporting in the evening, when in the fly ftate among the grafs and
herbage.

The Caterpillar becomes a Pupa about the beginning of June.—It
affixes itfelf by the tail to a ftalk of grafs in the fame manner as thofe
of the Butterfly genus, and like them is often found with the head
fufpended downwards ; it can by a fudden fpring turn itfelf upright
again.

In a little book entitled the AURELIAN's POCKET COMPANION,
by Mofes Harris, we find this fpecies defcribed, and called the *White
Plumed*, but the Linnæan fpecified Name *Didactyla* is added:—And
under the Linnæan name Pentadactyla (our prefent fpecimen) he has
defcribed the *Brown plumed* *.—The fame confufion is extended to
his folio work the AURELIAN. In Plate 1. he has figured the White
plumed under the fpecific name Didactyla, and in Plate 30, the Brown
plume, under Pentadactyla. Linnæus has comprifed all thofe Lepidopterous Infects whofe wings appear to confift of feveral diftinct
feathers, connected only at the fhafts, under the fubdivifion *Aluctæ*,
but Fabricius has given them the new name PTEROPHORUS, and
added the name Alucitæ to a fmall divifion of the Tinea, as Phal.
Chriftyloftella, &c. of *Linn.*

The Phal. Pentadactyla appears in the perfect ftate about the latter
end of June, fometimes earlier.

---

* Another fpecimen of the fame divifion of the genus (*Aluctæ*) but of a brown
colour " Alis fiffis fufcis, &c." *Linn.*

PLATE

PLATE CXI.

FIG. I. II.

CHRYSOMELA 4 PUNCTATA.

COLEOPTERA.

GENERIC CHARACTER.

Antennæ knotted enlarging towards the ends. Corfelet margined.

*** Body Cylindrical.

SPECIFIC CHARACTER.

Head and thorax black. Shells yellow brown with two black fpots on each. Antennæ ferrated.

CHRYSOMELA 4 *punctata* cylindrica, thorace nigro, elytris rubris : punctis duobus nigris, Antennis brevibus. *Linn. Syft. Nat.* 2. 374. 50. *edit.* 10.

CHRYSOMELA 4 *punctata* thorace nigro, elytris rubris, maculis duabus rubris antennis ferratis. *Degeer. Inf.* 5. 32. *tab.* 10. *fig.* 7.

Melontha coleoptris rubris maculis quatuor nigris, thorace nigro. *Geoff. Inf.* 1. 195. *tab.* 3. *fig.* 4.

Bupreftio 4 punctata. *Scop. Ent. Carn.* 206.

Cryptocephalus 4 punctatus. *Fab. Spec. Inf.* 1. 13³. 3.
Schœff. Elem. tab. 83. *fig.* 1.
——— *Icon. tab.* 6. *fig.* 1. 2. 3.
This

PLATE CXI.

This species is scarce, though more frequently met with than either of the following Chrysomelæ. It is generally found on the Hazel-nut tree.

FIG. III. IV.

CHRYSOMELA SANGUINOLENTA.

COLEOPTERA.

CHRYSOMELA.

SPECIFIC CHARACTER.

Black blue, a bright orange or red exterior margin to the elytra.

CHRYSOMELA *Sanguinolenta* ovata atra, elytris margine exteriori sanguineis. *Linn. Syst. Nat.* 2. 591. 38. *Syst. Ent.* 101. 40.

CHRYSOMELA nigro cœrulea, elytris atris punctatis margine exteriori rubro. *Geoff. Inf.* 1. 259. 8. *tab.* 4. *fig.* 8.

Chrysomela rubro marginata. &c. Degeer Inf. 5. 298. 7. *tab.* 8. *fig.* 26.

Buprestis Sanguinolenta. *Scop. carn.* 203.

Extremely rare in England; our specimen was found on the trunk of an ash tree in June 1794——n Kent.

FIG.

PLATE CXI.

9

FIG. V. VI.

CHRYSOMELA COCCINEA.

COLEOPTERA.

CHRYSOMELA.

SPECIFIC CHARACTER.

Fine red, with two black spots on each elytra, and one on the thorax.

CHRYSOMELA *coccinea* oblonga, thorace marginato fanguineo, macula nigra, elytris fanguineis maculis duabus nigris. *Linn. Syft. Nat.* 2. 592. 43.—*Fn. Sv.* 532.

CHRYSOMELA 4 *maculata*, &c. *Degeer Inf.* 5. 301. 10. *tab.* 9. *fig.* 1.

Coccinella Coleoptris rubris maculis 4 nigris. *Vdm. Diff.* 13. *Fab. Spec. Ent.* 1. 131. 83.

Very rarely met with : our fpecimen was taken on a thiftle in a field between Kennington Common and Camberwell, May 1794. The fpecies has not till very lately been confidered as a native of this country.

C PLATE

1

PLATE CXII.

SCARABÆUS FULLO.

COLEOPTERA.

GENERIC CHARACTER.

Antennæ clavated, their extremities fiffile. Five joints in each foot.

SPECIFIC CHARACTER,

AND

SYNONYMS.

Antennæ, of feven laminæ *. Head, thorax, and fhells brown, fpotted with white. Beneath white.

SCARABÆUS FULLO fcutellatus muticus, antennis heptaphyllis, corpore nigro pilis albis, fcutello macula duplici alba. *Linn. Syft. Nat.* 2. 553. 57.—*Fn. Sv.* 394.

SCARABÆUS, &c.   *Geoff. Inf.* 1. 69. 2.
       Frifch. Inf. 11. *tab.* 1. *fig.* 1.
SCARABÆUS Variegatus.   *Roef. Inf.* 4. *tab.* 30.
       Schæff. Icon. tab. 23. *fig.* 2.
       Hæfn. Inf. 2. *tab.* 7.
       Sulz. Hift. Inf. 1. 1.
Melolontha Fullo.   *Fab. Spec. Inf.* 1. 35. 1.

Except the Stag Beetle, (Cervus Lucanus) which is figured already in this work, this is the largeft Coleopterous Infect ever found in England; it is extremely rare, and is faid to be met with only in the fand on the fea coaft near Sandwich.

---

* The antennæ of the male is very large, as fhewn in our figure; the antennæ of the female is reprefented at **Fig. 1,**

PLATE CXIII.

FIG. I.

HEMEROBIUS HIRTUS,

NEUROPTERA.

Wings four, naked, tranſparent, reticulated with veins or nerves, Tail without a ſting.

GENERIC CHARACTER.

Mouth prominent. Palpi four. Wings deflexed. Antennæ longer than the thorax, taper, extended.

SPECIFIC CHARACTER.

Firſt wings tranſparent reticulated with brown veins, hairy. Veins fewer on the ſecond wings.

HEMEROBIUS hirtus, alis albis fuſco reticulatis, faſciis
duabus fuſceſcentibus. *Linn. Syſt. Nat.*
2. 912. 6.—*Fn. Sv.* 1507.
Degeer Inſ. 2. 2 70. 12. *tab.* 22. *fig.* 4. 5.

This very common Inſeꞓ is found on the nut tree, and oak. It conceals itſelf in the middle of the day among the foliage, or flies only in moiſt, ſhady places. It is always obſerved to be very briſk at the approach of a thunder ſtorm, like the Hemorobius Perla, &c.

D 2

The

PLATE CXIII.

The nerves on the wings are fo exceedingly delicate, that it is impoffible to give an accurate reprefentation of the natural fize; but to remedy that defect, we have fhewn the magnified appearance of an upper and under wing at Fig. 2.

The wings are of a pale tranfparent brown; which as the Infect moves in different directions reflect all the vivid colours of a Prifm.

PLATE

114

PLATE CXIV.

PHALÆNA COSSUS.

GOAT MOTH.

LEPIDOPTERA.

GENERIC CHARACTER.

Antennæ taper from the bafe.   Wings in general contracted when at reft.   Fly by night.

SPECIFIC CHARACTER,

AND

SYNONYMS-

Grey, with fhort black irregular curved lines on the upper wings. Antennæ feathered.

PHALÆNA COSSUS. *Bombyx* elinguis, alis deflexis nebulofis, thorace fafcia poftica atra. *Linn. Syft. Nat.* 2. 504. 40. *edit.* 10.

PHALÆNA pectinicornis elinguis, alis albo cinereis, ftriis tranfverfis nebulofis nigris. abdomine annulis albis.
> *Geoff Inf.* 2. 102. 4.
> *Degeer Inf. Vers. Germ.* 2. 1. 268. 1.
> *Merian. Europ tab.* 36.
> *Roef. Inf.* 1. *phal.* 2. *tab.* 18.
> *Reaum. Inf.* 1. *tab.* 17. *fig.* 1. 5.
> *Albin. Inf. tab.* 35. *fig.* 56.
> *Lyonet Traite de Chenille.*
> *Schœff. Icon. tab.* 61. *fig.* 1. 2.
> *Goed. Inf* 2. *tab.* 33.

The

The Caterpillar of the Goat Moth feeds on the internal fubftance of willow trees; it is faid to be alfo found in the body of the oak, but we have never difcovered any in fuch a fituation. The eggs are laid in the crevices of the trees; as foon as the Caterpillars are hatched, they begin to pierce into the folid wood. In moft parts of England they are called Auger Worms; the holes which they make in the timber appearing as if bored with that Inftrument.

It lives in the Caterpillar ftate three years before it is transformed to a pupa; when full fed it is four inches long, the body appears very flefhy, and without hairs; the head is black, and armed with very fharp forceps; the cafe is compofed of bits of wood and faw-duft, which it unites with a ftrong web; the infide is lined with a fine fmooth white filmy fubftance, like fattin; it paffes to the pupa ftate in the cavity which it has perforated in the caterpillar ftate, within three or four inches of the opening: it remains only two months in that ftate before the Fly is produced.

Is found in chryfalis in May; in the fly ftate, the latter end of June, or in July.

P L A T E

115

2

i

3

PLATE CXV.

CHRYSOMELA CEREALIS.

COLEOPTERA.

GENERIC CHARACTER.

Antennæ knotted, enlarging towards the ends. Corfelet margined.

SPECIFIC CHARACTER,

A N D

SYNONYMS.

Thorax and fhells ftriped with blue, crimfon and yellow green in-clining to gold. Wings fine fcarlet.

CHRYSOMELA CEREALIS. Ovata aurata, thorace lineis tribus coleop-trifque quinque cœruleis. *Syft. Ent.* 100. 33. *Linn. Syft. Nat.* 2. 588. 17.

CHRYSOMELA aurea fafciis cœruleis cupreifque alternis, punctis inor-dinatis.
Geoff. Inf. 1. 262. 14.
Schæff. Icon. tab. 1. *fig.* 3.
Fab. Spec. Inf. 1. *p.* 124. 45.

This beautiful Infect is a native of Georgia in North America; and has been received from feveral parts of Africa, as Guinea, &c.

E   It

It has alfo been found (though we believe very rarely) in the fouthern parts of Europe, particularly in Italy; and we have reafon to conclude it has been met with in the fouth of France, and in Germany *.

We prefume to include it among the Englifh Chryfomelæ, on the authority of the late Mr. Hudfon, author of the *Flora Anglica*, &c. who appears to be the only Naturalift that has taken it in Great Britain, except the Rev. Mr. Hugh Davies, of Beaumaris, who alfo met with a fpecimen of it on a mountain in Wales fome years fince.

The colour of the ftripes on the fhells fometimes vary; and the underfide, which in our Infect is purple, is often of a fhining brownifh colour; the tranfparent wings, which are concealed beneath the fhells, are bright red.

---

* Habitat in Europæ auftralioris fegete, in fpartio fcoparia.   *D. Prof. Hermann.* *Fab. Spe. Inf.*

PLATE

116

2

1

PLATE CXVI.

SPHINX CHRYSORRHŒA.

GOLDEN-TAIL SPHINX.

LEPIDOPTERA.

GENERIC CHARACTER.

Antennæ thickeſt in the middle. Wings, when at reſt, deflexed. Fly ſlow, morning and evening only.

SPECIFIC CHARACTER.

Wings tranſparent with black veins. Head, thorax, body, ſhining black with yellow rings or belts. Tail fine golden yellow.

––––––––––––––

In the paintings of *Ernſt*, a figure of a tranſparent-winged Sphinx, ſimilar to this, is given, under the ſpecific name Oeſtriformis: we are not clearly convinced he intended it for this Inſect, nor can we conceive the name to be by any means applicable; we therefore paſs over the reference to that very ſcarce work as doubtful, and reject his ſpecific name leſt he ſhould mean another Inſect.

Linnæus has not deſcribed this ſpecies, nor have we found a deſcription of it in the writings of Fabricius.

It is rare in England. THOMAS MARSHAM, Eſq. Sec. L. S. favoured me with the ſpecimen from which the annexed figure is taken; it was met with in Kenſington Gardens in June.

E 2   PLATE

117

PLATE CXVII.

PHALÆNA CRATÆGI.

OAK EGGER MOTH.

LEPIDOPTERA.

GENERIC CHARACTER.

Antennæ taper from the bafe. Wings, in general contracted when at reft. Fly by night.

SPECIFIC CHARACTER.

Wings rounded. Afh-colour, or dull brown, with obfcure waves of a darker colour.

PHALÆNA CRATÆGI.   *Linn. Syft. Nat.* 2. 823. 48.
                      Reaum. Inf. 1. tab. 44. *fig.* 10.
                      Degeer Inf. 1. tab. 11. *fig.* 20. 21.

We have never found this Infect common, though it muft not be confidered as a rare fpecies; it is feldom met with near London : our fpecimen was found in the Caterpillar ftate at Dartford in May. It changed to Chryfalis in June. The fly came forth in September.

The male is rather fmaller than the female generally, though not always. The ftrength of their colours is very inconftant, efpecially in the female, which we have feen very dark in fome fpecimens; in others nearly as pale as the male ; the general diftinction however between the two fexes is, the male being of a light grey with fpots and waves of brown, the female of an obfcure brown with fpots more diffufed.

PLATE

118

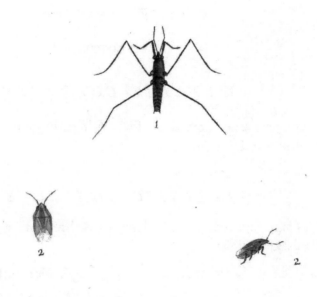

1

2

2

1

PLATE CXVIII.

FIG. I.

CIMEX LACUSTRIS.

HEMIPTERA.

Shells, or upper Wings, femi-cruftaceous, not divided by a ftraight future, but incumbent on each other. Beak curved downward.

GENERIC CHARACTER.

Antennæ longer than the Thorax. Thorax margined. In each Foot three joints.

SPECIFIC CHARACTER,

AND

SYNONYMS.

Above black. Beneath black changeable to white. Antennæ black, of four joints, half as long as the body. Eyes large, prominent. Fore Legs much fhorter than the reft.

Cinex Lacuftris. linearis niger, pedibus anticis breviffimis. *Linn. Syft. Nat.* 2. 732. 117.—*Fab. Spec. Inf. Fn. Sv.* 970.
Infectum Tipula dictum. *Bauh. Ball.* 213. *fig.* I.

This Infect is met with in great plenty on ftill waters, in fummer; it runs quick on the furface.

F   PLATE

PLATE CXVIII.

FIG. II.

CIMEX ACUMINATUS.

SPECIFIC CHARACTER,

AND

SYNONYMS.

Oval. Olive colour. Antennæ of five joints. Snout fharp. Thorax narrow before. Two brown longitudinal lines from the Eyes to the pofterior margin of the Target.

Cimex Acuminatus, &c.—*Linn. Syft. Nat.* 2. 723. 59.—*Fn. Sv.* 939.
Degeer Inf. 3. 271. 16. *tab.* 14. *fig.* 12, 13.
Mufca cimiciformis.   *Raj. Inf.* 56. 6.

———————————

Met with in *May*, on the Fern *. We have never found it common.

———————————

* Ofmunda Regalis.

PLATE

PLATE CXIX.

PHALÆNA ZICZAC.

Pebble Prominent Moth.

LEPIDOPTERA.

GENERIC CHARACTER.

Antennæ taper from the base. Wings in general contracted when at rest. Fly by night.

SPECIFIC CHARACTER,

AND

SYNONYMS.

Brown and white clouded like an Agate; a large clouded Eye, next to the exterior margin of the first Wings; on the interior margin a tuft, or appendage. Antennæ feathered.

Phalæna Ziczac. B. Alis deflexis dorso dentatis apicibusque macula grisea subocellari, antennis squamatis.

Syst. Ent. 573. 35. Linn. Syst. Nat. 2. 827. 61.—Fn. Sv. 1116.

Geoff. Inf. 2. 124. 29.

Merian. Europ. tab. 147.

Frisch. Inf. 3. tab. 1. fig. 2.

Degeer Inf. 1. tab. 6. fig. 1. 10.

Reaum. Inf. 2. tab. 22. fig. 9—16.

Fab. Spec. Inf. 2. p. 186. 76.

This singular and beautiful Caterpillar is found on the Willow, early in *June*; it becomes a Pupa within a fine, br   th web, which it spins between two or three leaves, (as represented in our Plate,) late in the same month; the Moth comes forth in *August*.

The

PLATE CXIX.

The trivial name prominent has been given to this Infect, becaufe when the Moth is at reft the remarkable appendages on the interior margin of the upper Wings form a prominent tuft above the back; we have fix different fpecies of Phalæna in this country which have the fame character, and are known among Collectors by the feveral names, Pale, Maple, Swallow, Iron, Pebble, and Cockfcomb, Prominents; the laft is common, the reft are generally very rare.

PLATE

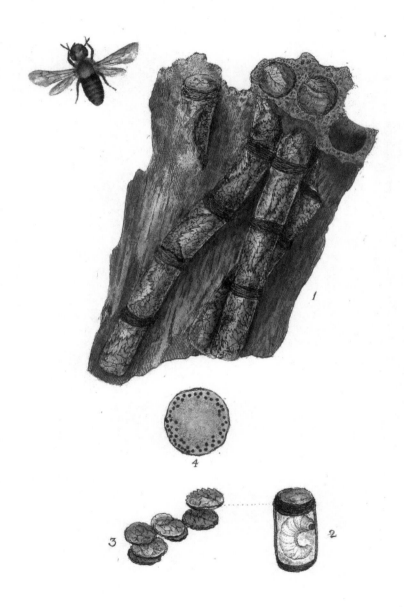

120

PLATE CXX.

APIS CENTUNCULARIS,

CARPENTER BEE.

HYMENOPTERA.

Wings four, generally membraneous. Tail of the females armed with a sting.

GENERIC CHARACTER.

Jaws, with a Trunk deflexed. Antennæ elbowed in the middle, first joint longest. Wings plain. Body hairy.

SPECIFIC CHARACTER,

AND

SYNONYMS.

Black. Body long, narrow. Head, Thorax, and Legs covered with greyish hair. Abdomen smooth, beneath covered with tawny hair.

Apis Centuncularis, nigra, ventre lana fulva.—*Syst. Ent.* 385. 42.—
    Linn. Syst. Nat. 2. 575. 4. edit. 10.
    Geoff. Inf. 2. 410. 5.
    Scop. carn. 799.
    Reaum. Inf. 6. tab. 10. fig. 3, 4.
    Fab. Spec. Inf. 1. 486. 59.

The wonderful instinct that directs the smallest Insects to provide for the safety of their future progeny, never fails to strike the attention of the inquisitive researcher into their oeconomy.—To perpetuate their

race

race is the great end of their being, and the moft aftonifhing effort of their ingenuity and care is employed to perfect this grand defign. We not only find innumerable eggs, and larvæ of Infect on all kinds of plants; in all ftanding waters; and in animal matter, when putrid; but many which can only be hatched from the egg by the warmth of living animals; thus the Tabanus pierces the thick hide of the Cow, and plunges its eggs into the flefh; the heat and moifture of which nourifhes both in the egg, and larva; the Hippobofca equina protrudes its eggs into the *rectum* of Horfes; and the Ichneumon into living Caterpillars: to thofe we could add many remarkable inftances of Infects, who have fhewn a lower fpecies of perception, by depofiting their eggs in places where the larvæ would find abundance of proper food; and with fuch ingenious contrivances for their fafety in a defencelefs ftate, as we could only expect from the fagacity of larger animals; but it is only our intention to premife with thofe general remarks, left the fubject we have chofen for our prefent Plate fhould be confidered as a folitary example of fuch ingenuity, and care towards their future offspring.

The Natural Hiftory of the Common Bee has been both fully and ably treated of, by *Schirach*, *Maraldi*, *Reaumur*, *Debraw*, and other authors of refpectability, and may be fuppofed to be pretty generally known by thofe converfant in rural affairs; the manners, however, of other fpecies of the fame genus has neither been fo fully explained, nor examined; they yet prefent a fund for the enquiries of the Naturalift, equally worthy his attention; though lefs beneficial; as the honey they make cannot be converted to our ufe.

Among the folitary Bees, fome penetrate into the earth, fcoop out hollow cavities; then polifh the fides within, and depofit their eggs, with proper food for the larvæ, till it becomes a Pupa. Others form nefts of loofe fand, which they glue together with a ftrong cement; thofe nefts are generally formed againft walls that are expofed to the fouth; without, they are rude and irregular, but within are very neatly finifhed, and divided into feveral cells or apartments, in each of which the Parent Bee lodges an egg. Of our prefent, and a few other fpecies, we may fay,

PLATE CXX.

33

" In firmeft oak they fcoop a fpacious tomb,
" And lay their embryo in the fpurious womb *."

We find this feafon, the Apis Centuncularis has done confiderable injury among the Timber Plantations in *Effex*; and we have fimilar information from fome parts of *Cambridgefhire*. A Gentleman fent me (early in the Spring) a piece of Oak, containing a quantity of the larva, from his plantation at *Birdbrook*, in *Effex*. He informs me, feveral Gentlemen in his neighbourhood had found large trunks of apparently healthy Oaks, completely perforated and filled with the larva of this mifchievous Infect; in many inftances the trunk had been materially injured, and the cafes were arranged as fhewn by the horizontal Section at Fig. 4, in our Plate.—The perforations were in a longitudinal direction, feveral feet through the folid timber, and when the leaves were frefh, appeared as fhewn at Fig. 1.

The Infect commences its operation at the upper part of the trunk of the tree; then boring in an oblique direction for about two inches or more, it follows a longitudinal courfe, it divides the ligneous fibres, or threads, till it forms the diameter of the cavity, which is about three-eighths of an inch, its depth various; fometimes only a few inches, at others, confiderably more; when the cavity is entirely formed, and all the duft and fragments cleared away, it finifhes the fides perfectly fmooth; the hardeft knot in the timber being infufficient to refift the ftrength of its jaws.—The cavity, when finifhed, appears divided by flight ridges, placed at the diftance of about three quarters of an inch from each other; this ferves to regulate the fize of each apartment or cell; and it now only remains to be lined for the reception of the egg: this lining is generally compofed of rofe-leaves; and is applied to the apartments in a very curious manner: the Parent Bee flies with a leaf to the orifice of the perforation, where fhe clips it round to the fize of the hole; this is forced to the bottom of the loweft cell; about feven, eight, or ten of fuch pieces form the firft layer; it next forms the fides, or cylindrical part of the lining; this is done by laying feveral whole leaves partly over each

---

* Brookes on Univerfal Beauty.

other,

race is the great end of their being, and the moſt aſtoniſhing effort of their ingenuity and care is employed to perfect this grand deſign. We not only find innumerable eggs, and larvæ of Inſect on all kinds of plants; in all ſtanding waters; and in animal matter, when putrid; but many which can only be hatched from the egg by the warmth of living animals; thus the Tabanus pierces the thick hide of the Cow, and plunges its eggs into the fleſh; the heat and moiſture of which nouriſhes both in the egg, and larva; the Hippoboſca equina protrudes its eggs into the *rectum* of Horſes; and the Ichneumon into living Caterpillars: to thoſe we could add many remarkable inſtances of Inſects, who have ſhewn a lower ſpecies of perception, by depoſiting their eggs in places where the larvæ would find abundance of proper food; and with ſuch ingenious contrivances for their ſafety in a defenceleſs ſtate, as we could only expect from the ſagacity of larger animals; but it is only our intention to premiſe with thoſe general remarks, left the ſubject we have choſen for our preſent Plate ſhould be conſidered as a ſolitary example of ſuch ingenuity, and care towards their future offspring.

The Natural Hiſtory of the Common Bee has been both fully and ably treated of, by *Schirach*, *Maraldi*, *Reaumur*, *Debraw*, and other authors of reſpectability, and may be ſuppoſed to be pretty generally known by thoſe converſant in rural affairs; the manners, however, of other ſpecies of the ſame genus has neither been ſo fully explained, nor examined; they yet preſent a fund for the enquiries of the Naturaliſt, equally worthy his attention; though leſs beneficial; as the honey they make cannot be converted to our uſe.

Among the ſolitary Bees, ſome penetrate into the earth, ſcoop out hollow cavities; then poliſh the ſides within, and depoſit their eggs, with proper food for the larvæ, till it becomes a Pupa. Others form neſts of looſe ſand, which they glue together with a ſtrong cement; thoſe neſts are generally formed againſt walls that are expoſed to the ſouth; without, they are rude and irregular, but within are very neatly finiſhed, and divided into ſeveral cells or apartments, in each of which the Parent Bee lodges an egg. Of our preſent, and a few other ſpecies, we may ſay,

6

" In

" In firmeſt oak they ſcoop a ſpacious tomb,
" And lay their embryo in the ſpurious womb *."

We find this ſeaſon, the Apis Centuncularis has done conſiderable injury among the Timber Plantations in *Eſſex*; and we have ſimilar information from ſome parts of *Cambridgeſhire*. A Gentleman ſent me (early in the Spring) a piece of Oak, containing a quantity of the larva, from his plantation at *Birdbrook*, in *Eſſex*. He informs me, ſeveral Gentlemen in his neighbourhood had found large trunks of apparently healthy Oaks, completely perforated and filled with the larva of this miſchievous Inſect ; in many inſtances the trunk had been materially injured, and the caſes were arranged as ſhewn by the horizontal Section at Fig. 4, in our Plate.—The perforations were in a longitudinal direction, ſeveral feet through the ſolid timber, and when the leaves were freſh, appeared as ſhewn at Fig. 1.

The Inſect commences its operation at the upper part of the trunk of the tree; then boring in an oblique direction for about two inches or more, it follows a longitudinal courſe, it divides the ligneous fibres, or threads, till it forms the diameter of the cavity, which is about three-eighths of an inch, its depth various ; ſometimes only a few inches, at others, conſiderably more ; when the cavity is entirely formed, and all the duſt and fragments cleared away, it finiſhes the ſides perfectly ſmooth; the hardeſt knot in the timber being inſufficient to reſiſt the ſtrength of its jaws.—The cavity, when finiſhed, appears divided by ſlight ridges, placed at the diſtance of about three quarters of an inch from each other; this ſerves to regulate the ſize of each apartment or cell; and it now only remains to be lined for the reception of the egg : this lining is generally compoſed of roſe-leaves; and is applied to the apartments in a very curious manner : the Parent Bee flies with a leaf to the orifice of the perforation, where ſhe clips it round to the ſize of the hole; this is forced to the bottom of the loweſt cell; about ſeven, eight, or ten of ſuch pieces form the firſt layer; it next forms the ſides, or cylindrical part of the lining; this is done by laying ſeveral whole leaves partly over each

---

* Brookes on Univerſal Beauty.

other,

other, as fhewn in our Plate, and cementing them together with a glutinous fubftance; thus the fides and bottom, each confifting of feveral layers, being finifhed, (in the form of a thimble) the Bee partly fills it with a kind of pafte, then throws over it a fmall quantity of leaves, reduced to powder, and depofits the egg; the covering to the whole is formed of the fame materials, and in the fame manner as the bottom; when fhe has forced about ten or fifteen circular pieces of leaves into the avenue and cemented them to the top, the covering is completed, and the egg is completely fecured from accident.—The covering feparated is fhewn in the Plate, at fig. 3, the larvæ, at fig. 2.

In this manner fhe proceeds with, and finifhes every cell diftinctly, till the perforation is entirely filled : in fome trees forty or fifty fuch perforations are placed within a quarter of an inch of each other.— The Bee comes forth late in Auguft; if the loweft is formed before thofe above, it eats its way up the channel, through their cafes.

Mr. *Adams*, in his Effay on the Microfcope, mentions a remarkable circumftance of a Bee (we fufpect of this fpecies). " A friend of mine (fays he) had a piece of wood cut from a ftrong poft * that fupported the roof of a cart-houfe, full of thefe cells or round holes, three-eighths of an inch diameter, and about three-fourths deep, each of which was filled with thefe rofe-leaf cafes, finely covered in at top and bottom."

---

* We learn this poft was fir.

PLATE

- 472 -

PLATE CXXI.

CURCULIO.

COLEOPTERA.

Wings two. Covered by two Shells, divided by a longitudinal suture.

GENERIC CHARACTER.

Antennæ clavated, elbowed in the middle, and fixed in the snout, which is prominent. Joints in each foot four.

FIG. I. II.

CURCULIO ÆQUATUS.

SPECIFIC CHARACTER.

Rostrum long, slender, dark brown sprinkled over with bronze; Thorax the same. Shells reddish brown. Legs brown.

Fab. Ent. Syst.

FIG. I. Natural Size.

This Insect was found in *May* on the hazel; the species varies in size, but more in colour.

G

PLATE CXXI.

FIG. III. IV.

CURCULIO PYRI.

SPECIFIC CHARACTER,

AND

SYNONYMS.

Snout ſhort. Thighs dentated. General colour bronze changeable to yellow red, brown, green, &c. Shells ſtriated and punctured.

CURCULIO PYRI. breviroſtris femoribus dentatis æneo fuſcus.

Linn. Syſt. Nat. 2. 615. 72.

Fn. Sv. 623.

Curculio breviroſtris, antennis fractis rufis, corpore oblongo æneo
nitido, pedibus rufis *. *Degeer Inſ.* 5. 246. 34.

Curculio viridis opacus, pedibus antenniſque magis fuſcis. *Linn. It.*
Scan. 355.

It has been ſuſpected by ſome Entomologiſts, that this Inſect ſhould only be conſidered as a variety of Curculio Argentatus. Much of its beautiful appearance depends on the time we take it in; when firſt hatched its colours are very rich and highly gloſſed with gold, but it gradually becomes dirty brown, or almoſt black.

The cauſe of this alteration in its appearance is eaſily perceived by the microſcope; the firſt, or ground colour is dark brown, but is entirely covered with oblong ſcales of various beautiful colours, particularly of a reddiſh gold, or bronze, interſperſed with thoſe of green, and brown colour; when the Inſect is firſt hatched, the ſcales lay over each other ſo as to conceal the ground colour; but as they rub off, or are otherwiſe injured, the brown becomes the general colour.——They vary alſo very much from red, to

---

* Variat pedibus rufis et nigris. *Fab. Spec. Inſ.* 1. 198. 217.

yellow,

PLATE CXXI.

37

yellow, or green hues, when firſt hatched; and are ſometimes found late in the ſeaſon, with almoſt every ſcale rubbed off.

Linnæus and Fabricius ſay, it is found on Pear trees †. We have met with it on ſeveral other trees. Found from *May* to *September*.

FIG. V. VI. VII.

CURCULIO CAPREÆ.

SPECIFIC CHARACTER.

Small, black. A longitudinal whitiſh line down the Thorax. Two waved white lines acroſs the ſhells, with a longitudinal mark of brown on each. Legs black.

CURCULIO CAPREÆ. *Fab. Spec. Inf.* 1. 168. 39.

This little Inſect very much reſembles Curculio Salicis, both in ſize and colours; but it is ſufficiently diſtinguiſhed from that ſpecies by its walking or running; as leaping is a particular character of that Curculio.

We have never met with more than one ſpecimen; found on the Ozier in *June*.

It is a very beautiful ſubject for the Opaque Microſcope; its magnified appearance is ſhewn at fig. 6.—The roſtrum fig. 7.—Fig. 5. Natural ſize.

---

† Habitat in Pyri foliis, in Corrolis declaratus. *Linn.*

G 2 PLATE

PLATE CXXI.

FIG. III. IV.

CURCULIO PYRI.

SPECIFIC CHARACTER,

AND

SYNONYMS.

Snout fhort. Thighs dentated. General colour bronze change-able to yellow red, brown, green, &c. Shells ftriated and punctured.

CURCULIO PYRI. breviroftris femoribus dentatis æneo fufcus.
Linn. Syft. Nat. 2. 615. 72.
Fn. Sv. 623.

Curculio breviroftris, antennis fractis rufis, corpore oblongo æneo nitido, pedibus rufis *. *Degeer Inf.* 5. 246. 34.

Curculio viridis opacus, pedibus antennifque magis fufcis. *Linn. It.*
Scan. 355.

It has been fufpected by fome Entomologifts, that this Infect fhoul only be confidered as a variety of Curculio Argentatus. Much of it beautiful appearance depends on the time we take it in ; when fir: hatched its colours are very rich and highly gloffed with gold, but gradually becomes dirty brown, or almoft black.

The caufe of this alteration in its appearance is eafily perceiv by the microfcope ; the firft, or ground colour is dark brown, b is entirely covered with oblong fcales of various beautiful colou particularly of a reddifh gold, or bronze, interfperfed with thofe green, and brown colour; when the Infect is firft hatched, t fcales lay over each other fo as to conceal the ground colou but as they rub off, or are otherwife injured, the brown becom the general colour.——They vary alfo very much from red,

----

* Variat pedibus rufis et nigris. *Fab. Spec. Inf.* 1. 198. 217.

yell

PLATE CXXI. 37

yellow, or green hues, when first hatched; and are sometimes found late in the season, with almost every scale rubbed off.

Linnæus and Fabricius say, it is found on Pear trees †. We have met with it on several other trees. Found from *May* to *September*.

FIG. V. VI. VII.

CURCULIO CAPREÆ.

SPECIFIC CHARACTER.

Small, black. A longitudinal whitish line down the Thorax. Two waved white lines across the shells, with a longitudinal mark of brown on each. Legs black.

CURCULIO CAPREÆ. *Fab. Spec. Inf.* 1. 168. 39.

This little Insect very much resembles Curculio Salicis, both in size and colours; but it is sufficiently distinguished from that species by its walking or running; as leaping is a particular character of that Curculio.

We have never met with more than one specimen; found on the Ozier in *June.*

It is a very beautiful subject for the Opaque Microscope; its magnified appearance is shewn at fig. 6.—The rostrum fig. 7.—Fig. 5. Natural size.

† Habitat in Pyri foliis, in Corrolis declaratus. *Linn.*

G 2

PLATE

122

PLATE CXXII.

SPHINX ELPENOR.

ELEPHANT SPHINX, or HAWK-MOTH.

LEPIDOPTERA.

GENERIC CHARACTER.

Antennæ thickeft in the middle. Wings, when at reft, deflexed. Fly flow, Morning and Evening.

SPECIFIC CHARACTER,

AND

SYNONYMS.

Wings angular, entire; firft wings ftriped tranfverfely with greenifh brown, and red. Second Wings red, with a white pofterior margin; black at the bafe. Body red and brown.

SPHINX ELPENOR. Alis integris, viridi purpureoque variis, pofticis rubris bafi atris.
> Fab. Spec. Inf. 2. 148. 43.
> Syft. Ent. 543. 25.
> Linn. Syft. Nat. 2. 801. 17.
> Fn. Sv. 1049.

Sphinx fpirilinguis, alis viridi purpureoque fafciatis, fafciis linearibus tranfverfis. Geof. Inf. 2. 86. 10.
> Roef. Inf. 1. phal. 2. tab. 33. fig. 73.
> Petiv. Gazoph. tab. 40. fig. 11. 12. 17.
> Frifch. Inf. 13. 4. tab. 2.

The Caterpillars of this very elegant Sphinx are generally found in marfhy places in *June* and *July*. They feed on the Convolvulus, Vine,

Vine, and fome other plants, but prefer white ladies bedftraw; they caft their fkins feveral times, and when full fed are fome green, and others of a brown colour. The Caterpillars of the female is a fine green elegantly marked with black, as reprefented in our plate; thofe of the male are varied with the fame dark markings, but the colour is a dull brown inclining to black in thofe parts where the females are green.

It poffeffes a faculty peculiar to a very few Infects, it can protrude its head and three firft joints to a tapering point; or entirely conceal the head and contract the firft joints, by drawing them apparently into its body.

The Caterpillars form a white fpinning among the leaves in *Auguft*; remains in the pupa ftate during the winter; the Fly comes forth *May* following. They are frequently deftroyed by an Ichneumon fly.

P L A T E

123

PLATE CXXIII.

CIMEX PRASINUS.

HEMIPTERA.

Shells, or Upper Wings femicruftaceous, not divided by a ftraight future, but incumbent on each other. Back curved downwards.

SPECIFIC CHARACTER,

AND

SYNONYMS.

Head, Corfelet and Shells green. Abdomen black above, with a yellow and black margin, beneath pale orange varied into green. Legs and Antennæ yellowiſh.

Cimex PRASINUS. *Linn. Syſt. Nat.* 2. 722. 49.
                 Fab. Spec. Inf. 2. 354. 96.

Not uncommon in the month of *Auguſt* in woods. Found on the Oak.

PLATE

124

PLATE CXXIV.

PHALÆNA ANOSTOMOSIS.

SCARCE CHOCOLATE-TIP MOTH.

LEPIDOPTERA.

GENERIC CHARACTER.

Antennæ taper from the bafe. Wings in general deflexed when. at reft. Fly by night.

BOMBYX.

Antennæ feathered.

SPECIFIC CHARACTER.

Firft wings greyifh, with three tranverfe ftripes of dull white. Apex fine chocolate colour. Second wings and body pale brown.

PHALÆNA ANOSTOMOSIS.   B. alis deflexis grifeis, ftrigis tribus albidis fubanaftomofantibus, thorace ferruginato. *Fab. Spec. Inf.* 2. 189. 85.

Linn. Syft. Nat. 2. 824. 53.

Fn. Sv. 1124.

Goed. Inf. 1. *tab.* 33.

A very rare fpecies of Phalæna. In the perfect ftate it is feldom met with ; and in the Caterpillar ftate few Collectors are acquainted with its haunts. It feeds on the fallow, willow, and poplar, and may be found fometimes by ftripping off the bark of thofe trees.

H   Our

Our specimen was taken in the vicinity of Oak-of-Honor Hill, Surry. The Caterpillar was met with when it was ready to spin its web, in which state it is represented; its spinning was formed between the folds of a leaf in the month of October, the Moth came forth in May.

The Moth in the upper part of the plate is a small specimen of the female; it differs very little from the male, except that the antennæ of the latter is much feathered, as is shewn on the back of the leaf.

The species is more plentiful on the continent of Europe, and a variety of it is a native of some parts of North America.

A Collector of Insects in London met with a brood of this species last September, in the Caterpillar state, containing more than twenty; some were covered with a milk-white down, others inclining to grey, but in general they were like the specimen given in our plate. They changed their appearance frequently, and some were much larger than the rest. The Moths also differ very much both in size and colour; some are dingy, others have the chocolate colour much diffused; and in general, when the Insect is perfect, it is beautifully varied with a pale bloom of a purple hue.

PLATE

125

PLATE CXXV.

MUSCA SEMINATIONIS.

DIPTERA.

Wings 2.

GENERIC CHARACTER.

A foft flexible trunk, with lateral lips at the end. No palpi.

SPECIFIC CHARACTER.

Head and Thorax black-brown; Abdomen black, with very minute fpecks of white. Wings clouded and fpeckled with brown. A yellow ftreak on the under fide of the abdomen.

MUSCA SEMINATIONIS.　　Antennis fetariis, alis atris cinereo punc-
　　　　　　　　　　　　tatis, abdomine bafi fubtus flavo.
　　　　　　　　　　　　Fab. Spec. Inf. 2. 452. 90.

This fpecies is fometimes met with in meadows, on plaintain, thiftles, &c. in May and June.

It is a very pleafing object for the Microfcope, particularly the wings, which are finely reticulated and fpotted. Its magnified appearance is given with its natural fize in our plate.

H 2　　　　　　　　　　　PLATE

126

PLATE CXXVI.

PHALÆNA RUMICIS.

BRAMBLE MOTH.

LEPIDOPTERA.

GENERIC CHARACTER.

Antennæ taper from the bafe. Wings in general contracted when at reft. Fly by night.

* NOCTUA.

Antennæ like a hair.

SPECIFIC CHARACTER.

Firft wings grey, marked with pale black ftreaks and clouds, with an eye in the middle, and two white fpots on the anterior margin. Second wings pale brown.

PHALÆNA RUMICIS.   N. criftata, alis deflexis cinereo fufcoque variis litura marginis tenuioris alba.— *Fab. Spec. Inf.* 2. 238. 143.

PHALÆNA RUMICIS.   fpirilinguis criftata, alis deflexis cinereo bimaculatis, litura marginis tenuioris alba. *Linn. Syft. Nat.* 2. 852. 164.— *Fn. Sv.* 1200.
Merian. Europ. tab. 82.
Alb. Inf. tab. 32.
Wilk. pap. 26. *tab.* 3. *a.* 1.
Degeer Inf. 4. *tab.* 9. *fig.* 2.

The Caterpillar of this Moth are ufually found on the Bramble, from which it has received its Englifh name; it is not, however, wholly confined to that food, as we have fed it on grafs and other plants indifcriminately put into its breeding-cage. It paffes to the chryfalis ftate in September; the Fly appears in May.

PLATE

127

PLATE CXXVII.

BUPRESTIS SALICIS.

COLEOPTERA.

GENERIC CHARACTER.

Antennæ taper, the length of the thorax: Head half concealed.

SPECIFIC CHARACTER.

Head and thorax fine blue.   Shells upper half changeable green ; lower part reddiſh purple.

BUPRESTIS SALICIS: elytris integerrimis viridis nitens, coleopteris aureis baſi viridibus.   *Fab. Gen. Inſ. Mant. p.* 237.

Bupreſtis elegantula, Schrank. Inſ. Auſtr. n. 365. *p.* 195.

Cucuius rubinus. Fourcroy. Ent. Paris. T. I. n. 4. *p.* 33.

Le Richard rubis. Geoff. Inſ. Paris F. I. p. 126.

Geputzter Stinkkäfer. Weiden-Prachtkäfer. *Panz. Faun. Inſ. Germ.*

––––––––––––

This uncommonly beautiful Inſect has been deſcribed as a native of Algiers in Africa, a figure of it is given in the work of *Olivier,* and another in Panzer's Hiſtory of the Inſects of Germany ; we find it alſo deſcribed by *Fourcroy* and *Geoffrory* as a native of France, but cannot learn that it has been conſidered as an Engliſh ſpecies before this time.

We were not ſo fortunate as to take this Inſect, it was communicated by a perſon on whoſe veracity we can rely : he found it on the bark of an old willow tree, between Dulwich Common and

I   Norwood,

Norwood, on, or about the 8th of June, 1794. As we know the precife fpot where it was taken, we fhall attend to it particularly next feafon, and the earlieft intimation of fuccefs, fhall be given through the medium of a future number.

When we confider how much the ftudy of coleopterous Infects has been neglected in this country, even by thofe who have purfued with unremitting perfeverance almoft every other branch of Entomology, we cannot be much aftonifhed that fuch a minute Infect as the *Bupreftis Salicis* fhould have efcaped notice ; add to this, we can fcarcely doubt that it is very rare in this country, and probably lives concealed in the crevices of the tree, or under the rotten part of the bark. The number of new Infects that have been difcovered in this country within a few years *, renders it not improbable, that future Englifh Entomologifts, by extending their enquiries, may find many more of the fpecies that are now met with in the northern, and perhaps even fouthern parts of Europe.

Fig. 1, natural fize.   Fig. 2, its magnified appearance.

---

* Among the rarities met with laft fummer, we may particularly mention the Phalæna Delphinii, *Peafe Bloffom Moth*. This very beautiful Infect was taken by a Gentleman at Chelfea ; it was never afcertained before to be an Englifh Infect.

P L A T E

128

P L A T E CXXVIII.

EPHEMERA VULGATA.

COMMON EPHEMERA, or MAY-FLY.

NEUROPTERA.

Wings 4. Naked, tranſparent, reticulated with veins or nerves. Tail without a ſting.

GENERIC CHARACTER.

Antennæ very ſhort. Two protuberances before the eyes. Wings erect. Second pair ſmall. Two or three tails like briſtles. Short lived.

SPECIFIC CHARACTER.

Wings reticulated, browniſh with five or ſix brown ſpots. Body yellowiſh, with black ſpecks. Three tails.

EPHEMERA VULGATA: cauda triſeta, alis nebuloſo maculatis.
Linn. Syſt. Nat. 2. 906. 1. *Fn. Sv.* 1472.

In the larva and pupa ſtate, this Inſect is found under looſe ſtones at the bottom of ſhallow pools ; in the winged ſtate it frequents the water.

We have ſeveral ſpecies of this genus in England. The Ephemera Vulgata, is the largeſt among them. A very diſtinguiſhing character of them is the ſhortneſs of their lives, which ſeldom exceeds a few hours. In the month of May theſe Inſects are ſeen in great plenty on the water, where they are greedily devoured by the fiſh ; anglers ſay, when the large Ephemera appears, the trout will ſnap at no other bait, than the artificial fly made after its form.— In ſome ſpecimens the wings are more clouded, and the tails longer than in others.

I 2　　　　　　　　　P L A T E

129

PLATE CXXIX.

FIG. I, II, III.

PHALÆNA HASTATA.

ARGENT AND SABLE MOTH.

LEPIDOPTERA.

GENERIC CHARACTER.

Antennæ taper from the bafe. Wings in general contracted when at reft. Fly by night.

* * *GEOMETRÆ.*

SPECIFIC CHARACTER

AND

SYNONYMS.

White, beautifully marked and fpotted with black.

PHALÆNA HASTATA: feticornis, alis omnibus nigris albo maculatis, fafciis duabus albis nigro punctatis haftata dentatis. *Linn. Syft. Nat.* 2. 870. 254. *Fn. Sv.* 1276.

Phalæna antennis filiformibus; alis latis albis fafciis undulatis maculifque haftatis nigris. *Degeer. Inf. Verf. Germ.* 2. 1. 334. 7. *tab.* 8. *fig.* 20. *Clerk. phal. tab.* 1. *fig.* 9. *Kleman Inf.* 1. *tab.* 44.

The Argent and Sable Moth is fcarce in the Fly ftate : though its young caterpillars are not uncommon in fome parts of Kent ; we

have

have met with several about the narrow lanes in Darent-wood, Dartford, in April, or early in the month of May. It is however very difficult to breed them ; they generally die in the pupa state, or before they cast their last skin when caterpillars ; from several specimens taken during the three last summers, we have only had one Moth produced, and that so crippled, as merely to enable us to ascertain the species.

The small Caterpillars are of a dark purplish colour, when nearly full fed they have a yellow under side marked with black, with the back purple ; before they change to the pupa state, they become almost brown,

They remain only a month in the pupa state. The Moth appears about the middle of June.—Food, white-thorn and alder.

P H A L Æ N A A N A S T O M O S I S.

F I G. IV.

Since the publication of the last Number, we have been favoured with a most beautiful specimen of the Moth figured in the 124th Plate of this Work, and present a figure of it to our subscribers, together with the several changes of the Phalæna Hastata ; it will shew how very liable this Insect is to variation in its colours, size, &c.

We find also that though this Insect has been named Phalæna Anastomosis in the most scientific Cabinets in London, and always received as such by the best authority, it is not the Insect referred to by Fabricius in his Species Insectorum under that title ; that Author, as well as Linnæus, refers under the specific name *Ph. Curtula* to the 43d Plate of the third Volume of Roesel's Insects ; in this Plate is figured a Moth which is certainly a species distinct from our Insect, and is well known by its Linnæan name *Curtula,* or English title *Chocolate Tip* ; yet Fabricius gives an additional reference for the same species to the 11th Plate of Roesel's

fourth

PLATE CXXIX.

57

fourth Volume of Infects, and in this we find the figure of a Moth whofe markings and general appearance correfpond with our fpecimen, though its colours are totally different, being a very pale grey with fcarcely any dark colour near the apex of the wings ; the larva much more refembles our figure, and induces us to conclude, that though the figure of this laft Moth is fo extremely different, it is probably intended for the fame fpecies as our Infect ; and therefore that the two diftinct fpecies have been confounded together, by a falfe quotation of Linnæus's Amanuenfis.

And we are partly confirmed in this fuppofition by the words of Linnæus himfelf; he fays, *Ph. Anaftomofis* is very like *Ph. Curtula,* but the Moth figured in Rœfel's plate, and referred to in the Synonyms under Anaftomofis, does not bear the leaft refemblance to it ; —our Infect on the contrary, though evidently a diftinct fpecies, is not unlike it.

PLATE

130

PLATE CXXX.

GRYLLUS VIRIDISSIMUS.

HEMIPTERA.

Shells, or upper Wings, femicruftaceous, not divided by a ftraight future, but incumbent on each other. Beak curved down.

GENERIC CHARACTER.

Head maxillous, and with palpi. Antennæ filiform. Wings folded. Hind Legs ftrong, for leaping.

SPECIFIC CHARACTER

AND

SYNONYMS.

Head, Thorax, and Wings green, without fpots. Antennæ very long.

GRYLLUS VIRIDISSIMUS: thorace rotundato, alis viridibus immaculatis, antennis fetaceis longiffimis. *Linn. Syft. Nat. v.* 1. *p.* 430. 38. *edit.* 10.

LOCUSTA VIRIDISSIMA: alis viridibus immaculatis, antennis longiffimis. *Fab. Syft. Ent.* 286. 22.— *Spec. Inf.* 1. 359. 23.

Locufta viridis cantatrix viridis immaculata, thorace rotundato, cauda feminæ enfifera recta. *Deg:er Inf.* 3. 428.

Agrigoncus. Lift. Goed. 301. *tab.* 121.

K   This

This Insect is larger than the great green Grafshopper, (*Gryllus verrucivorus*) or any other species of the genus we have in this country ; unlefs we notice the *Gryllus Magratorius*, which is well known for its depredations in many parts of the world, but is rarely met with in England.

The prefent species is perhaps not uncommon in many places, but it is very difficult to difcover its hiding-places in the day-time ; its chirp is fometimes heard in a calm Summer's evening, about fun-fet, iffuing from the bufhes where it is concealed ; and from which it feldom ventures till night ; it continues its chirping at intervals till morning.

The female feems to prefer a warm, and rather moift fituation, to depofit her eggs in, and this is commonly the fide of a bank that is expofed to the fun ; but is well covered with grafs and other herbage to keep it moift. She is furnifhed with a fharp double edged fheath, like a *fword*, with which fhe opens the ground in a perpendicular direction ; firft fcooping out a convenient cylindrical aperture, and then widening the lower part into a fpacious apartment for the reception of the eggs. *See* Fig. I.

When the Insect burfts from the egg it is very minute, and without Wings ; in this ftate it nips the tender fhoots of grafs, &c. It foon increafes in fize and affumes the pupa form ; in which ftate though the Wings are not perfect, their rudiments appear next the Thorax : it continues in this ftate till it has nearly acquired its full fize before the Wings burft open from the protuberances.

Our fpecimens were taken in Batterfea Meadows ; in the egg ftate early in April ; winged ftate in June.

PLATE

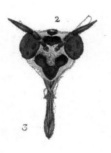

131

PLATE CXXXI.

TABANUS CAECUTIENS.

DIPTERA.

Wings two.

GENERIC CHARACTER.

Antennæ conic, of four fegments. Trunks flefhy, terminated by two lips. Palpi, one on each fide of the Trunk.

SPECIFIC CHARACTER

AND

SYNONYMS.

Eyes brilliant, green with black fpots. Thorax brown with yellowifh lines. Body bright yellow with triangular black marks, anterior margin, and center of the Wing black.

TABANUS CAECUTIENS : oculis viridibus nigro punctatis, alis maculatis. *Fab. Syft. Ent. n.* 18. *p.* 790.
Fab. Spec. Inf. 2. 27. *p.* 459.

TABANUS CAECUTIENS : oculis nigro-punctatis, alis maculatis.—
Linn. Syft. Nat. 17. *p.* 1001. *ed.* 13.
n. 17. *p.* 2885.
Faun. Suec. n. 1888.

Tabanus fufcus, abdominis lateribus pedibufque flavis, alis maculis
fufcis. *Geoffr. Inf.* 2. *n.* 8. *p.* 463.

Tabanus nibulofis. Harris Inf. Angl. tab. 7. *fig.* 5.

Mufca bipennis pulcra, alis maculis amplis albis pictis. *Rai. Inf.*
p. 272.

Le Taon brun, à cotes du ventre jaunes, et ailes tachetées de noir.
Geoffr.

Die buntaugichte Breme. *Panzer's Deutfchlands Infecten, &c.* —
Faun. Inf. Germ.

K 2

In

In the months of June and July, or earlier in forward feafons, this Infect is found in great plenty in the lanes and fkirts of woods; and are very troublefome to perfons or animals who pafs through fuch places in the middle of the day: they conceal themfelves in the crevices of the bark of trees, or among the foliage till about an hour before noon, when they come forth in great plenty, and fettle on the hands and face, or other thinly covered parts, and dart their fharp pointed trunks or probofcis into the flefh: we have obferved the fting of this Infect to be moft fevere about mid-day, particularly when the fun fhines bright, and emits much heat; a difagreeable fenfation continues in the ftung part for fome time, and is generally fucceeded by a large tumor, and a flight difcharge of pungent fluid before it difappears entirely.

To explain more minutely the ftructure of the trunk, we have given a figure of its magnified appearance at Fig. III: the outer coat, or fheath, terminates at its extremity, in two lateral, moveable lips, and contains a longitudinal, horny, convex blade; the acute point of which is concealed between thefe lips: this interior tube, when examined with a Microfcope, appears to confift of three others, exceedingly fharp at the points; and are ufed by the Infect as lancets to lacerate the flefh when it feeds, while it pumps, or fucks up the blood and moifture from the wound, through the capillary tubes with which feveral parts of the trunk are furnifhed.

It feeds in the fame manner on Infects, but chiefly on thofe in the larva ftate.

The fpecies is not uncommon in Germany, France, Italy, and moft warm parts of Europe: alfo received from Georgia, in North America.

Fig. I. Natural Size. Fig. II. Front View of the Head magnified, with the Eyes and Probofcis; the former are moft beautiful microfcopical objects when the Infect is alive, but turn brown after it dies.

PLATE

132

PLATE CXXXII.

PHALÆNA LUNARIA.

BEAUTIFUL THORN-MOTH.

LEPIDOPTERA.

GENERIC CHARACTER.

Antennæ taper from the bafe. Wings, in general deflexed when at reft. Fly by night.

SPECIFIC CHARACTER.

Antennæ feathered. Wings angulated, indented; a *lunar* fpot near the center of each. General colour, pale red brown, clouded and fpeckled.

Kleman Inf. 3.
Fab. Spec. Inf. 2. 245. 18 ?

We have been furnifhed with the larva of this extremely rare Infect in a fingular manner: a wafted fpecimen of the female was taken in the Summer of the year 1794, and depofited a quantity of eggs in the box in which it was ftuck; thefe hatching fome time after, a great number of young Caterpillars were produced; feveral of a full fize, paffed to the pupa ftate, (in a reddifh web fpun on the leaves) and four fine Moths came forth laft Summer.

The eggs were very minute, perfectly globular, and of a pale greenifh colour: the clufter confifted of more than feventy, and few of them proved abortive; but fome of the largeft Caterpillars devoured

voured the reft, and many others wandered from the food, and fo perifhed. The Moths were far fuperior for the beauty and richnefs of their colours to any fpecimens we have feen before; but this is not remarkable, as moft of the fpecimens preferved in Cabinets near London, have been taken in the winged ftate.

It feeds on the lime; is found in the Caterpillar ftate in Auguft; the Moth appears in June. Is fometimes taken in the interior part of Darent-Wood, Dartford; and rarely elfewhere near London. It has been met with alfo at Feverfham, in Kent, on the Elm.

Kleman, in the laft volume of German Infects, lately publifhed, has given three figures of this Moth; but has neither figured the Caterpillar nor Pupa. *Fabricius* is the only fyftematical Writer who appears to have noticed it.

PLATE

133

PLATE CXXXIII.

PHALÆNA PSI.

GREY DAGGER-MOTH.

LEPIDOPTERA.

GENERIC CHARACTER.

Antennæ taper from the bafe. Wings, in general deflexed when at reft. Fly by night.

Nóctua, Antennæ fetaceous.

SPECIFIC CHARACTER
AND
SYNONYMS.

Firft wings and body grey; on the former three or four black marks, in the form of a dagger*. Second wings pale brown, with a flightly fcalloped margin.

PHALÆNA PSI: criftata, alis deflexis cinereis, anticis lineola bafeos
charaĉteribufque nigris. *Syft. Ent.* 614. 104.—
Fab. Spec. Inf. 2. 235. 129.
Linn. Syft. Nat. 2. 846. 135.
Alb. Inf. tab. 26.

———————

The Caterpillars of the Grey Dagger-Moth is frequently found on fruit trees; particularly on the cherry: it feeds alfo on the willow and poplar, and on almoft all plants indifcriminately when confined in the breeding cage. It is not an uncommon Infeĉt: the Caterpillars change in September, remain in the chryfalis ftate during winter, and the Moth appears late in May, or early in June.

————————————

* Or like the Greek (ψ) *Pfi*; from which it receives its fpecific name.

L   PLATE

134

PLATE CXXXIV.

PHALÆNA PLANTAGINIS.

SMALL TIGER.

LEPIDOPTERA.

GENERIC CHARACTER.

Antennæ taper from the bafe. Wings in general deflexed when at reft. Fly by night.

Bombyx antennæ of the male pectinated or feathered.

SPECIFIC CHARACTER.

Firft Wings yellow, fecond Wings orange colour; both clouded with black. Body orange and black.

PHALÆNA PLANTAGINIS elinguis. alis deflexis atris, rivulis flavis, inferioribus rubro maculatis. *Linn. Syft. Nat.* 2. 820. 42.—*Fn. Sv.* 1132.

PHALÆNA pectinicornis elinguis, alis deflexis, fuperioribus fufcis, maculis luteis, inferioribus rubris, maculis quatuor nigris. *Geof. Inf.* 2. 109. 10.

Phalæna Alpicola. Scop. carn. 507.
Wilk. pap. 24. *tab.* 3. *a.* 5.
Roef. Inf. 4. *tab.* 24.
Fab. fpec. Inf. 2. 196. 115.

L'Ecaille brune. *Geofr.*

Der Wegerichfpinner. Die fpanifche Fahne. Die befchleierte Bärenphalene. *Panf. Fauz. Inf. Germ.*

L 2   This

This fpecies feeds on nettles, chickweed, plantain, grafs, &c.
The Caterpillars very much refemble thofe of the large *Garden Tiger
Moth* *, except in fize ; they change into chryfalis about the middle
of April, and appear in the winged ftate the latter end of May.

We have not found this Infect fo plenty as the *Ruby Tiger* Moth †,
and it is infinitely more fcarce than the great *Garden Tiger* Moth,
figured in the early part of this Work.

A variety of this fpecies, with crimfon under wings, is found in
the Eaft Indies and in America. The under wings of the female,
in the European fpecimens, are much redder than in the male.

---

* Phal. Caja.   † Phal. Fuliginofa.

PLATE

135

PLATE CXXXV.

CIMEX SPISSICORNIS.

HEMIPTERA.

GENERIC CHARACTER.

Antennæ longer than the thorax. Thorax margined. In each foot three joints.

SPECIFIC CHARACTER

AND

SYNONYMS.

Antennæ very large. Head, thorax, and shells, pale blackish brown. Feet yellow.

CIMEX SPISCICORNIS: oblongus niger, pedibus flavis, antennis incraffatis. *Fabri. Gen. Inf. Mant. p.* 300.—*Sp. Inf.* 2. 207. *p.* 372.
Die borstenhornige Wanze. *Panz. Inf. Germ.*

The singular structure of the antennæ of this minute Insect, recommends it to particular notice. They are nearly as long as the body, and in the thickest part are very bulky; hence it has received the specific name Spissicornis, or large horned Cimex.

It is not uncommon in summer; flies amongst bushes or low herbage in the day time: the lower wings are of a very beautiful purple colour, and give a blackish hue to the outer wings when
folded

folded under them. The larva we fufpect has not been figured, if noticed, before, and for this reafon we have given it of the natural fize at fig. 1. and its magnified appearance at fig. 2.—at fig. 3. the natural fize of the perfect or winged infect; fig. 4. the fame magnified.

Found in the larva ftate in May, was fed on grafs, the winged Infect appeared June 19th.

P L A T E

136

PLATE CXXXVI.

PHALÆNA HEXADACTYLA.

MANY-FEATHERED MOTH.

LEPIDOPTERA.

GENERIC CHARACTER.

Antennæ taper from the bafe. Wings in general deflexed when at reft. Fly by night.

* PTEROPHORUS.

SPECIFIC CHARACTER.

Wings divided into Feathers, yellowifh and grey, with brown Spots.

Phalæna Hexadactyla. *Linn. Syft. Nat.*
PHALÆNA HEXADACTYLUS, alis fiffis cinereis, fingulis fexpartitis.
      Fab. Spec. Iuf. 2. 312. 7.—*Syft. Ent.*
      672. 7.
      Reaum. Inf. I. *tab.* 19.—*Fig.* 19. 21.
      Frifch. Inf. 7. *tab.* 73.

Among an almoft endlefs variety of fpecies, which the tribes of Infects prefent, few have a more fingular appearance than the little creature we have felected for our prefent fubject. It is perhaps one of the moft curious pieces of natural mechanifm (if we may be allowed the expreffion) that can be conceived, for of a moft complicated fabric which the wings appear, every part, though feparate,

---

* *Fabricius.*

M   acts

acts in perfect unifon with the reft ; in moft winged Infects we find the tendons of each wing united by ftrong membranaceous webs, which prevent any one from acting without the others, but in this every tendon muft perform a diftinct part, and yet perfectly in conformity with the reft to affift the Infect in its flight. When the Infect refts the feathers fold over one another ; but when it flies, they are thrown open, and refemble a full expanded fan.

The natural fize of this fingular creature is given at Fig. 1. and as a more correct figure than can be fhewn in fuch a fmall compafs was thought neceffary, its magnified appearance is reprefented at Fig. 2.

The plumes of this Infect differs fo much from thofe of other Moths, that we have alfo added, at Fig. 3, the appearance of the upper part of one, as feen by a very deep lens of the Microfcope ; by this the ftem or quil is obferved covered with fcales of the form ufually found on other Moths, but the fides are finely feathered with long hairs, in tufts, alternately of a light and dark colour, and which, owing to the minutenefs of the Infect appear like patches of an uniform colour, before it is examined with the Microfcope.

This Infect is not uncommon in Summer, it flies about hedges in the evening.

PLATE

137

PLATE CXXXVII.

PHALÆNA CHRYSITIS.

BURNISHED BRASS MOTH.

LEPIDOPTERA.

GENERIC CHARACTER.

Antennæ taper from the Bafe. Wings in general deflexed when at reft. Fly by night.

NOCTUA.

Antennæ of both fexes filiform.

SPECIFIC CHARACTER

AND

SYNONYMS.

Firft Wings brown, with two tranfverfe broad waves of greenifh gold on each. Second Wings blackifh grey. Wings margined.

PHALÆNA CHRYSITIS *Linn. Syft. Nat.* 2. 843. 126.
　　　Noctua criftata, alis deflexis orichalceis, margine fafciaque
　　　　grifeis. *Syft. Ent.* 606. 69.—*Spec. Inf.* 2. *p.* 226. 91.
　　　　—*Fabricius.*
Phalæna feticornis fpirilinguis, alis deflexis ferrugineo fufcis, fafcia
　　　　duplici tranfverfa viridi aurea. *Geof. Inf.* 2. 149. 97.
Phalæna antennis filiformibus, dorfo criftato, alis deflexis grifeis,
　　　　fafciis duabus aureo viridibus. *Degeer. Inf. Vers. Germ.*
　　　　2. 1. 311. 2.
　　　　Merian. Europ. tab. 39.
　　　　Albin Inf. tab. 71. *fig. a. b. c. d.*
　　　　Schæff. Icon. tab. 101. *fig.* 2. 3.

M 2　　　　　　　　　　　The

The pencil can produce but a feeble and inadequate imitation of the metallic fplendour of this beautiful, yet common Infect. The upper Wings have the appearance of fine burnifhed brafs, changeable in different directions of the light to green, brown, and rich golden hues ; the under Wings are of a blackifh colour, and ferve as an admirable contraft to the more brilliant and varied teints of the upper Wings. The Thorax is crefted.

Berkenhout has given a very falfe defcription of the Caterpillar of this Infect, he fays it is " fmooth, orange with white fpots ;" we think it neceffary to note this error only as it may miflead young Collectors, who have no other affiftant than his Synopfis, by which they can determine the Species, when in the Caterpillar ftate. It is aftonifhing how he could poffibly be led into this error, when *Albin,* *Fabricius* *, and all preceding authors on Entomology, have defcribed it fo plainly.

It feeds on Nettles, and other Plants, growing among the low herbage by the fide of banks ; in fine feafons there are generally two broods of them from May, to June in the following year ; the firft are found early in May in the Caterpillar ftate, appear in June in the winged ftate ; Caterpillars are full fed again in July, the Moths come forth in Auguft.

---

* *Larva* folitaria, gibbofa , viridis albo ftriata. *Fabricius.*

P L A T E

138

PLATE CXXXVIII.

FIG. I. II. III.

CASSIDA NOBILIS.

COLEOPTERA.

Wings two, covered by two fhells, divided by a longitudinal future.

GENERIC CHARACTER.

Antennæ knotted, enlarging towards the ends.   Shells and Thorax bordered.   Head concealed under the corfelet.

SPECIFIC CHARACTER.

Greyifh Green; on the center of each Shell a ftreak of gold, which dies with the Infect.   Body beneath black.

CASSIDA NOBILIS: grifea elytris linea cœrulea nitidiffima,
Linn. Syft. Nat. 2. 575. 4.
Oliv. Inf. 97. *tab.* 2. *fig.* 24.
Raj. Inf. 107. 7.

This fpecies is far lefs common than *Caffida Viridis.*   It is a very beautiful Infect; but, like moft other minute fpecies, appears with infinitely more advantage in the Microfcope for opake objects; indeed, without fuch affiftance, it is impoffible to perceive the beauty of that part by which it is diftinguifhed from every other fpecies of the fame genus we have in England, the lines of fine gold and blue, which are feen on the middle of the Shells.

When the Infect is alive, it is of a pale greenifh colour, inclining to brownifh grey, and along the middle of each Shell appears a fplendid ftreak, or line of gold, margined with a fine pale fky blue, alternately varying into green, and gold.   By the Microfcope we

alfo

alfo difcover many minute punctures, and feveral waved lines and ftreaks, which defcend along the Shells from the bafe, and unite near the apex.

Its colours are more or lefs beautiful as the Infect is healthy or fickly ; and as it dies, the colours gradually perifh ; the fplendor of gold is no longer vifible than life is retained, it changes to green ; from green to a brown, which fcarcely appears through a faint tinge of blue, and in a few hours it changes altogether to a rufty brown colour.

This Infect is admirably protected from external injury by the fingular form of its Thorax and Shells, which are alfo fo large as to conceal every other part when the Infect walks.

The natural fize is fhown at fig. 4, (upper fide.) Fig. 2, under fide. Fig. 3, upper fide magnified.

F I G. IV.

CHRYSOMELA BANKII.

COLEOPTERA.

GENERIC CHARACTER.

Antennæ knotted, enlarging towards the ends. Corfelet margined.

SPECIFIC CHARACTER

Body oval. Head, Thorax and Shells, purplifh olive colour, changeable, with a bronze appearance. Beneath, reddifh brown, or teftaceous.

CHRYSOMELA BANKII : ovata fupra ænea fubtus teftacea. *Fab. Entomologia Syftematica. T.* 1. 310. 16.

This is a very rare Infect in England. It refembles *Chryfomela bicolor* in fize, and colour of the Head, Thorax and Shells ; but it may be readily diftinguifhed from that fpecies by the teftaceous colour of the under fide, that part being wholly of a violaceous colour in *C. bicolor.*

Found in May on a thiftle.

F I G.

9

PLATE CXXXVIII. **81**

FIG. V. VI.

CICADA DILATATA.

HEMIPTERA.

Shells or upper Wings, femi cruftaceous, divided by an oblique future, and incumbent on each other. Beak bent down.

GENERIC CHARACTER.

Antennæ taper. Shells membraneous. In each foot three joints. Hind legs ftrong for leaping.

SPECIFIC CHARACTER.

Entirely brown, pale with faint whitifh and dark lines, a fmall black fpot on the center of each wing.

A figure of this Infect is given in Villers's Entomology as a native of France; in this he follows the authority of Fourcroy, who has a defcription of the fame fpecies in his Catalogue of Infects, found in the environs of Paris. This laft author calls it Le Cigale renflée, from its puffed or fwelled appearance. The name given by Villers is Cicada dilatata.

The confufion made by Fabricius, in his alterations of the *Linnæan genera*, renders it doubtful whether he has defcribed this Infect, though, from its being commonly found in moft parts of Europe, we muft fuppofe he has not paffed over it without notice: we have examined his laft work, (Entomologia Syftematica, &c.) and cannot find an Infect anfwering our fpecies with any reference either to Fourcroy or Villers, we therefore prefer the fpecific name given by the latter author.

Is found in June; and is lefs common than any Infect of the fame genus hitherto given in this work.

PLATE

139

PLATE CXXXIX.

PHALÆNA METICULOSA.

ANGLE-SHADES MOTH.

LEPIDOPTERA.

GENERIC CHARACTER.

Antennæ taper from the bafe. Wings, in general deflexed when at reft. Fly by night.

* *Noctua* antennæ like briftles in both fexes.

SPECIFIC CHARACTER

AND

SYNONYMS.

Firft Wings pale reddifh colour, with a broad triangular brown fpake in the middle. Second Wings palifh, with dark waves; margin of both Wings indented.

Phalæna Meticulofa. *Linn. Syft. Nat.* 2. 845. 132.—*Fn. Sv.* 1164.

Phalæna Meticulofa: alis deflexis, erofo dentatis, pallidis, anticis bafi incarnata, triangule fufco. *Fab. Syft. Ent.* 608. 78.

Phalæna feticornis fpirilinguis, alis deflexis margine erofis cinereo fufcis, fuperioribus triangulo marginali fuf-cefcente, incarnatum includente, thorace gibbo. *Geof. Inf.* 2. 151. 84.
Merian. Europ. tab. 24.
Albin Inf. tab. 13.
Roef. Inf. 4. *tab.* 9.
Degeer Inf. 1. *tab.* 5. *fig.* 14.
Goed, Inf. 1. *tab.* 56.

N   The

The Phalæna Meticulofa certainly exceeds many other Infects of the fame tribe for elegance and fimplicity: the variety of teints fo delicately, indeed almoft infenfibly foftened into one another, and neatnefs of the waves and lines interfperfed over the whole, amply compenfate for the defection of more gaudy colours. In the cater-pillar ftate it is fcarcely lefs deferving attention; the yellow fpecks on a beautiful, yet lucid green, have a very pleafing effect. The web it fpins round its pupa is of a fine white colour, and filky tex-ture; the pupa within of a blackifh chocolate colour.

This fpecies is fometimes met with in plenty, though lefs fo in fome feafons than in others; and not unfrequently is more abundant when the feafon appears moft unfavourable. It feeds on nettles chiefly, but we have found it on feveral other plants; and once on a young oak, in Kent; the leaves of which we fed it on fome time. In the caterpillar ftate it is found in April, changes to the pupa ftate in May, the Moth appears in June.

P L A T E

140

PLATE CXL.

SCARABÆUS FASCIATUS.

YELLOW BEETLE.

COLEOPTERA.

Wings two, covered by two shells, divided by a longitudinal suture.

GENERIC CHARACTER.

Antennæ clavated, their extremities fissile. Five joints in each foot.

SPECIFIC CHARACTER

AND

SYNONYMS.

Head, Body, Thorax, black : covered with long, yellowish hairs. Shells pale yellow, with three transverse black stripes on each. Abdomen longer than the Shells.

SCARABÆUS FASCIATUS scutellatus muticus niger tomentoso flavus, elytris fasciis duabus luteis coadunatis. *Linn. Syst. Nat.* 2. 556. 70. *Fn. Sv.* 395.

TRICHIUS *fasciatus:* niger tomentoso flavus, elytris fasciis tribus nigris abbreviatis. *Fab. Syst. Ent.* 40. 1. —*Spec. Inf.* 1. 48. Nº I.

Scarabæus niger hirfuto flavus, elytris luteis, fasciis tribus nigris interruptis. *Geoff. Inf.* 1. 80. 16.

N 2 *Drury*

PLATE CXL.

Drury Inf. 1. *tab.* 36. *fig.* 2.
Degeer. Inf. 4. *tab.* 10. *fig.* 19.
Voet. Scar. tab. 5. *fig.* 43.

In Germany this Insect is not uncommon: we believe it is very rare in this country. Found generally on umbelliferous plants.

PLATE

141

PLATE CXLI.

PHALÆNA DOMINULA.

SCARLET TIGER MOTH.

LEPIDOPTERA.

GENERIC CHARACTER.

Antennæ taper from the bafe. Wings, in general contracted when at reft. Fly by night.

Bombyx antennæ of Male feathered, Female fetaceous.

SPECIFIC CHARACTER

AND

SYNONYMS.

Firft Wings black gloffy green, with orange and white fpots. Second Wings and Abdomen fcarlet, with black fpots.

Phalæna Dominula: alis incumbentibus atris, maculis albo flavef-centibus, pofticis rubris nigro maculatis.
Fab. Syft. Ent. 583. 93.—*Spec. Inf.* 2. 200. 130.

Phalæna Dominula. *Noctua* fpirilinguis lævis, alis depreffis nigris: fuperioribus cæruleo flavo aloque, inferioribus rubro maculatis. *Linn. Syft. Nat.* 2. 509. 68 *edit.* 10.

Formerly this beautiful Moth was found in great abundance at *Charlton* in *Kent*, but within the laft two or three years moft of the

broods

broods have been wantonly deftroyed, and they are now feldom met with. In the caterpillar ftate they feed on nettles and hound's-tongue *, changes to the pupa ftate about the middle of May, and in June the Moth comes forth.

---

* *Cynogloffum officinale.*

PLATE

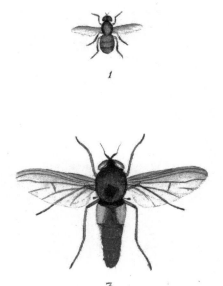

PLATE CXLII.

FIG. I.

MUSCA AURATA.

DIPTERA.

Wings two.

GENERIC CHARACTER.

A foft flexible trunk, with lateral lips at the end, no Palpi.

SPECIFIC CHARACTER

AND

SYNONYMS.

Head brown. Thorax polifhed, greenifh, or braffy. Abdomen flat, obtufe, brownifh gold-colour. Legs yellowifh ; Feet brown.

Mufca aurata : antennis fetariis nitida thorace æneo, abdomine obtufo aureo. Fabricius. *Ent. Syft. Vol. VI.* 335. 37.——*Mantiffa. Vol. II. p.* 347. *No.* 63.

This Infect has been only noticed in the latter writings of Fabricius. We have not found it uncommon in the fummer upon the leaves of Fruit trees ; and particularly on fuch as grow againft a fouth wall : they fly brifkly about noon, when the fun fhines.

O   FIG.

PLATE CXLII.

FIG. II. III.

MUSCA SEMI-ARGENTATA.

SPECIFIC CHARACTER.

Eyes brown. Thorax green; changeable to filver. Abdomen filvery, with fhades of bright yellow, and grey, and fome tranfverfe ftreaks of black, very changeable.

Mufca femi-argentata. *Marfham's MSS.*

We do not find that this rare and beautiful Infect has been defcribed either by *Linnæus* or *Fabricius*. It was taken a few years fince in Epping Foreft by Mr. Bentley, an eminent Collector of Englifh Infects, and noticed by Thomas Marfham, Efq. Sec. L. S. in his Manufcript Notes, under the fpecific name Semi-argentata. Several fpecimens of it were taken laft June in Epping Foreft; except them, we have not heard of any being met with for fome time.

Fig. 2. natural fize.   Fig. 3. magnified.

PLATE

143

PLATE CXLIII.

PAPILIO ARGUS.

COMMON BLUE BUTTERFLY.

LEPIDOPTERA.

GENERIC CHARACTER.

Antennæ knobbed at the end. Wings, when at reft, erect. Fly by day.

SPECIFIC CHARACTER.

Male upper fide fine blue with white margins. Female dark brown, with a patch of blue on the middle of each wing. Underfide of both fexes lightifh brown, with black and red fpots.

Papilio Argus : alis ecaudatis, pofticis fubtus limbo ferrugineo ocellis coeruleo argenteis. *Fab. Syft. Ent.* 525, 346.— *Linn. Syft. Nat.* 2. 789. 232. *Fn. Sv.* 1074. *Roef. Inf.* 3. *tab.* 37. *fig.* 3—5. *De Geer Inf.* 4. *f.* 14. 15. *Wilk. Pap.* 63. *t.* 1. *a.* 1. *Merian. Europ. tab.* 153. *Schæff. Icon. tab.* 29. *fig.* 3. 4.

Though this beautiful Infect is very common in fome places in the Butterfly ftate, we have never met with it's larva, nor with any account of it that appeared fatisfactory. In that ftate it feems fcarcely known. It is faid, by fome Collectors, to be a plain green Caterpillar, with very few hairs, bulky, and broadeft acrofs the middle. It certainly feeds very low among the thickeft grafs, or perhaps like

O 2   fome

fome larvæ of Moths, never comes above the furface of the ground, and lives on the roots of grafs.

The Male is of a fine blue colour on the upper fide, and elegantly marked on the under fide with white circles, having a black fpot in the center of each : the wings are alfo bordered with fimilar fpots, marked with a vermillion colour. The Female has very little appearance of the fine blue of the Male : the upper wings are of a dull brownifh black, with a bluifh colour on parts, and marked with a few red and black fpots : the underfide as in the Male.

They feem to delight in Meadows, and, like all other Butterflies, are on the wing only in the day time. The firft brood appears in the Fly ftate in June.

P L A T E

144

PLATE CXLIV.

PHALÆNA VIRIDANA.

SMALL GREEN OAK MOTH.

LEPIDOPTERA.

GENERIC CHARACTER,

Antennæ taper from the bafe. Wings, in general deflexed when at reft. Fly by night.

** Tortrix. *Linn.*

SPECIFIC CHARACTER

AND

SYNONYMS.

Firft Wings pea green. Second Wings dufky.

Phalæna viridana. Pyralis. Alis rhombeis, anticis viridibus immaculatis.—*Fabricius. Syft. Ent.* 656. 4.—*Linn.
Syft. Nat.* 2. 875. 266.

Phalæna feticornis fpirilinguis, humeris latis, antennis flavefcentibus,
alis dilute fufcis.—*Geof. Inf.* 2. 171. 123.
Reaum. Inf. 2. *tab.* 18. *fig.* 6. 7.
Roef. Inf. 1. *phal.* 4. *tab.* 1.
Frifch. Inf. 3. *tab.* 8.

Early in July we find this fpecies flying about the narrow paths
and lanes in woods where Oaks are plenty. It is obferved to fhelter
itfelf in the day time, generally among fuch trees as have the foliage
thick

thick and the bark covered with mofs, &c. and very feldom among young trees. In the Caterpillar ftate it lives concealed in a fine filky web, fpun up on the leaves. When it is difturbed it drops by a fingle thread from one branch to another, the glutinous fubftance of the thread adhering wherever it touches, fo that if it is damaged in any part the Infect is in no danger of falling, unlefs the laft faftening breaks off. The Caterpillar changes to the pupa ftate early in June : the firft appearance of the Moth is commonly about the end of the fame month.

In England we have another fmall Moth (Phalæna Chlorana) which at firft fight may be miftaken for Phalæna Viridana. It differs from this Infect in feveral refpects ; the under Wings are whiter, and the ftripe along the anterior margin of the upper Wings incline more to a cream colour than in our prefent fpecies ; the Caterpillar alfo is very different and feeds on the Willow.

LINNÆAN

LINNÆAN INDEX

TO

VOL. IV.

COLEOPTERA.

HEMIPTERA.

LEPIDOP-

INDEX.

LEPIDOPTERA.

NEUROPTERA.

* Not defcribed before.

HYMEMOP-

INDEX.

HYMEMOPTERA.

DIPTERA.

P.   ALPHA-

ALPHABETICAL INDEX

TO

VOL. IV.

INDEX.

ERRATUM to Vol. IV.

PLATE CXXIV. *for* Phalæna Anoftomofis, *read* Phalæna Anaftomofis.

THE

NATURAL HISTORY

OF

BRITISH INSECTS;

EXPLAINING THEM

IN THEIR SEVERAL STATES,

WITH THE PERIODS OF THEIR TRANSFORMATIONS,
THEIR FOOD, OECONOMY, &c.

TOGETHER WITH THE

HISTORY OF SUCH MINUTE INSECTS

AS REQUIRE INVESTIGATION BY THE MICROSCOPE.

THE WHOLE ILLUSTRATED BY

COLOURED FIGURES,

DESIGNED AND EXECUTED FROM LIVING SPECIMENS.

By E. DONOVAN.

VOL. V.

LONDON:

PRINTED FOR THE AUTHOR,

And for F. and C. RIVINGTON, N° 62, ST. PAUL's CHURCH-YARD.

MDCCXCVI.
S

ADDRESS TO SUBSCRIBERS.

THE Proprietors beg leave to apprize the Subscribers to this Work, that, in future, the Letter-press for two Numbers will be published together, but that the Plates will be sold in monthly Numbers as usual.——By this mode of publication the Subscribers will be in possession of the Descriptions before the Plates are published; and the Proprietors will avoid a very heavy Stamp-duty.

By an Act of the Legislature in the reign of Queen Anne *, a Duty was imposed on all *Pamphlets*, not exceeding One Sheet, or sixteen Pages, of Letter-press in Octavo; but as that regulation was probably intended only to repress the circulation of Pamphlets of an immoral or seditious † nature, the Commissioners of the Stamp-duties have never demanded this Duty, on periodical Publications of a Scientific Nature, though the Numbers of every Work are regularly entered at the Stamp-Office, and a Duty paid on them within fourteen days after their publication. The Commissioners have, however, now determined to exact this additional Duty, and have signified to all Publishers, that in future *a Duty must be paid on every Copy of every Pamphlet, not exceeding One Sheet*, as the Law directs; and farther that the Stamp must be impressed on the first page of every such Pamphlet, as on Almanacks, Newspapers, &c. We conceive therefore that it will be far better to adopt this mode of avoiding such Duty, and having every volume of our Work *disfigured* by twelve Stamps in the face of the Letter-press.

The NATURAL HISTORY of BRITISH BIRDS will be published in the same manner.

---

* 10 *Ann. c.* 19. *f.* 101.

† School-Books, Books of Devotion, Acts of Parliament, &c. are expressly exempted.

145

NATURAL HISTORY

OF

BRITISH INSECTS.

PLATE CXLV.

PAPILIO RHAMNI.

BRIMSTONE BUTTERFLY.

LEPIDOPTERA.

GENERIC CHARACTER.

Antennæ clavated, or knobbed at the end. Wings, when at reſt,
erect. Fly by day.

SPECIFIC CHARACTER

AND

SYNONYMS.

Wings angulated, entire, pale yellow, with a brown ſpot near the
center of each. Underſide very pale yellow. Antennæ reddiſh.

PAPILIO RHAMNI. Alis integerrimis angulatis flavis, ſingulis
puncto ferrugineo.—*Linn. Syſt. Nat.* 2. 765. 106.
—*Fn. Sv.* 1042.

A 2   *Papilio*

Papilio præcox fulphurea five flavo viridis, fingulis alis macula fer-
　　　ruginea notatis.——*Raj. Inf.* 112. 4.
　　　Sulz. Inf. tab. 13. *fig.* 84.
　　　Roef. Inf. 3. *tab.* 46. *fig.* 1. 2. 3.
　　　————— 4. *tab.* 26. *fig.* 1. 5.
　　　Degeer Inf. 1. *tab.* 15. *fig.* 1. 10.
　　　Efp. Pap. 1. *tab.* 4. *fig.* 4.
　　　Schæff. Elem. tab. 94. *fig.* 7.
　　　————— *Icon. tab.* 35. *fig.* 1. 3.

The Brimftone Butterfly is common in many places in the month
of June in the Fly ftate. In the Caterpillar ftate it is feldom
taken, and when in chryfalis it is generally concealed among the
herbage, where it is almoft impoffible to be difcovered. In this
ftate, like all other fpecies of the Butterfly tribe, it is fufpended by
the tail, but has fuch mufcular ftrength, that if touched it can throw
itfelf upright immediately, in the fame manner as the Chryfalis of
Phalæna pentadactyla. It feeds chiefly on buck-thorn, whence it has
received the fpecific name Rhamni.

P L A T E

146

1

3

2

PLATE CXLVI.

BOMBYLIUS MEDIUS.

DIPTERA.

Wings two.

GENERIC CHARACTER.

Trunk taper, very long, between two horizontal valves.

SPECIFIC CHARACTER

AND

SYNONYMS.

Thorax and body yellowifh brown, white at the extremity. Wings with brown fpots.

BOMBYLIUS MEDIUS: alis fufco punctatis corpore flavefcente pof-
tice albo.—*Linn. Syft. Nat.* 2. 1009. 2. 1919.
BOMBYLIUS PUNCTATUS niger villis fulvis, alis fufco punctatis.—
De Geer. Inf. 6. 269. 2. *tab.* 15. *fig.* 12.
Schæff. Elem. tab. 27. 1.
———— *Icon. tab.* 78. *fig.* 3.
Fab. Syft. Ent. 802. 2.
———— *Spec. Inf.* 2. 473.

The Bombylius genus is very concife. *Fabricius* in the *Species Infectorum* enumerates only nine fpecies, of thofe five are found in Europe, major, medius, minor, ater and fufcus; the three former are natives of this country; the fourth is frequent in Germany, the
last

laſt in Italy.—To theſe Fabricius has added a few ſpecies in his laſt
work *Entomologia Syſtema*, which have not been deſcribed before, but
they are all exotics.

The ſpecies figured in the annexed plate is not common. It lives
on the nectareous juice of flowers. Is found in May.

F I G.   III.

M U S C A   H Y P O L E O N.

DIPTERA.

GENERIC CHARACTER.

A ſoft flexible trunk, with lateral lips at the end. No palpi.

SPECIFIC CHARACTER.

Eyes brown. Thorax black, margined with yellow. Abdomen
black, with five yellow ſpots. Legs yellow.

Muſca Hypoleon. *Lin. Syſt. Nat.*
Stratiomys Hypoleon. *Fab. Mantiſa.* 2. *p.* 347. *N*° 63.

This Inſect was taken laſt Auguſt, flying among ſome ruſhes in
Batterſea meadows. The line at Fig. 2. denotes the natural ſize.

P L A T E

147

1

PLATE CXLVII.

GRYLLUS GRYLLOTALPA.

MOLE CRICKET.

HEMIPTERA.

Shells or upper wings femi-cruftaceous, not divided by a ftraight future, but incumbent on each other, beak curved down.

GENERIC CHARACTER.

Head maxillous, and with palpi. Antennæ filiform, or taper. Wings folded. Hind legs ftrong for leaping.

SPECIFIC CHARACTER

AND

SYNONYMS.

Dark brown. Antennæ filiform, long, fmall. Head long and fmall. Four fhort palpi. Corfelet cylindrical, fhells fmall, veined, wings long. Body hairy. Two fmall tails. Fore feet large, palmated.

GRYLLUS GRYLLOTALPA. *Linn. Syft. Nat.* 2. 693. 10.

Gryllus fupra fufcus, fubtus ferrugineo flavus, pedibus anticis latis, compreffis denticulatis. *De Geer. Inf.* 3. 517. 2.

Acheta gryllotaipa: alis caudatis elytro longioribus, pedibus anticis palmatis. *Fab. Syft. Ent.* 279. 1.— *Sp. Inf.* 1. 353. 91. 1.

Catefby Carol. 1. *tab.* 8.

Frifch. Inf. 11. *tab.* 5.

Seb. Muf. 4. *t.* 89. *fig.* 3. 4.

Sulz. Inf. tab. 9. *fig.* 59.

Roef. Inf. 2. *Gryll. tab.* 14. 15.

It

PLATE CXLVII.

It is scarcely possible to find a more singular creature than the Mole Cricket. It lives in burrows which it forms about an inch or more below the surface of the ground. The female deposits a large bed of eggs about the size of small pease, rather of an oval form, and brownish colour. They are laid in a circular cavity, which is two or three inches wide, and near an inch in height. An aperture is made on one side, with an easy ascent to the surface of the ground, and is ingeniously covered at the top with loose earth. When the young larvæ are first hatched, they scarcely exceed the twelfth of an inch in length. They ascend through the opening, and subsist on the plants nearest their habitation, till their fore claws have acquired sufficient strength to burrow into the earth. In the larva state they nearly equal the perfect Insect in size, and resemble it in every respect, except that they have no wings. The shells appear first; this is the pupa state, and shortly after the membraneous wings appear also. It makes very little use of its wings, as they are too weak to support its body long; and indeed it has not much occasion for them, as it lives in the same manner as the Mole, and, like it, is furnished with powerful claws, with which it can burrow through the ground to a very considerable distance.

This destructive creature is generally found in great numbers wherever they once deposit their eggs; for it is impossible to pursue and destroy them without doing much injury to the ground they infest. If they find a way into a kitchen-garden, they sometimes destroy whole beds of young plants in the space of one night; and this is not astonishing, when we consider that they seldom eat any part except the roots, which they nip very close, and consequently the other parts must perish. They seem particularly fond of Lettuces.

Fig. I. one of the fore claws.

PLATE

148

PLATE CXLVIII.

PHALÆNA POTATORIA.

DRINKER MOTH.

LEPIDOPTERA.

GENERIC CHARACTER.

Antennæ taper from the base.   Wings in general deflexed when at rest.   Fly by night.

BOMBYX.

Antennæ, male feathered, female, like a bristle.

SPECIFIC CHARACTER

AND

SYNONYMS.

Yellow brown.   Wings slightly scalloped; on each of the upper wings an oblique line, and two white spots near the anterior margin. Female paler colour than the male.

PHALÆNA POTATORIA: alis reverfis fubdentatis flavis, ftriga fulva repandaque, punctis duobus albis.—*Syst. Ent.* 564. 28.
PHALÆNA maxima alis e fulvo flavicantibus.   *Raj. Inf.* 143. 3.
        Goed. Inf. 1. *tab.* 12.
        Sepp. Inf. 4. 37. *tab.* 8.
        Schæff. Icon. tab. 67. *fig.* 10. 11.
        Wilk. pap. 27. *tab.* 3. *b.* 2.

B

The

The Caterpillars of this Infect feed on grafs, they are found in May, and the Moth appears about the middle of June.

The female differs in feveral refpects from the male; it is of a buff colour, and is generally, though not always, larger. The chryfalis is black, and is enclofed in a ftrong yellowifh cafe, as fhewn in the plate.

PLATE

149

PLATE CXLIX.

ATTELABUS CURCULIONOIDES.

COLEOPTERA.

Wings two, covered by two fhells, divided by a longitudinal future.

GENERIC CHARACTER.

Antennæ thicker towards the end.   Head narrow behind.   Four joints in each foot.

SPECIFIC CHARACTER

AND

SYNONYMS.

Shells and thorax red.   Head black.

ATTELABUS CURCULIONOIDES: niger thorace elytrifque rubis.—
Lin. Syft. Nat. 2. 619. 3.
Rhinomacer niger thorace elytrifque rubris, probofcide longitudine
capitis.—*Geof. Inf.* 1. 273. 10.
Curculio Nitens, *Paykull. Monogr.* 130. 122.
Schæff. Icon. tab. 56. *fig.* 7.
Sulz. Inf. tab. 4. *fig.* 12.

A pair of this very fingular and rare fpecies was taken on a young nut tree in Darent Wood, Dartford, early in May, 1795.

The remarkable ftructure of it's head deferves particular notice; it is fhaped like a vafe, and when the Infect is alive is protruded

B 2   far

far beyond the thorax by it's long flender neck. It has alfo a very bufy motion of it's head from the right to the left when it runs: we obferve a fimilar motion in many Infects; but as few have fuch a flender neck, it is feldom fo quick and repeated as in this.

The natural fize is given in the upper part of the plate, the magnified appearance of the head is fhewn below.

PLATE

150

PLATE CL.

FIG. I.

PHALÆNA MARGINATA.

LEPIDOPTERA.

GENERIC CHARACTER.

Antennæ taper from the bafe.   Wings in general deflexed when at reft.   Fly by night.

NOCTUA.

Antennæ fetaceous.

SPECIFIC CHARACTER

AND

SYNONYMS.

Upper wings, yellow brown, with four ftreaks of red brown acrofs each; two circles of the fame colour in the middle; fpace next the exterior margin dark colour.   Lower wings pale brown with a fpot of black in the center, and band of black next the pofterior edge.

NOCTUA MARGINATA: Criftata, alis deflexis flavefcentibus, ftrigis ferrugineis poftice fufcus. *Fabricius Spec. Inf.* 2. 230. 108.—Mant. *Inf.* 2. *p.* 166. *n.* 209.

Tabellar. Uerz. II. heft. p. 41. *n.* 59.   *Noctua rutilago* criftata, alis deflexis flavis, ferrugineo ftrigofis fafciaque poftica fufca; pofticis pallidis limbo nigro.

Berliner. Mag. 3. Stuct. p. 294. *n.* 41.

Phalæna Umbra. Die Zimmetmotte.

Gefenius handb. *p.* 162. *n.* 77.   *Phal. Noct. Umbra.* Die Zimmetmotte.

De VILLIERS ent. *Linn.* 2. *p.* 258. *n.* 280. *Phal. Noctua Marginata.* la Bordure.

Phalæna Marginata. Klemann's. Infecten Gefchichte, *&c.* Kurnberg; 1792. Vol. 2. *pl.* 7. *fig.* 6. 7. 8.

The

The Synonyms of this rare Infect have been more minutely collected, than is common in the defcriptions of this work, as it has been generally confidered an undefcribed fpecies. Mr. Crow, of Feverfham, who has enriched the collections of feveral gentlemen in London, with many curious Infects, met with two or three fpecimens of this Moth, and among others fent one to Mr. Bentley, a collector in London, a few years fince. I have to acknowledge being favoured with this Infect by Lord William Seymour; his Lordfhip met with it in Wiltfhire.

Fabricius, in the *Species Infectorum*, has made a very confiderable error; and which it is proper to notice in this place, " *Noctua Marginata*, native of America," and defcribed from the Collection of Dr. Hunter, occurs in page 216. fpec. 40; and again in page 230, fpec. 108, " *Noctua Marginata*, a native of Europe," the prefent fpecimen. The former he has indeed changed to " *Noctua Marginella*" in his laft work, *Entomologiæ Syftematicæ*, but without the flighteft notice of the firft miftake, or any reference to the *Species Infectorum*.

It appears to be a native of Germany by the laft work publifhed by *Klemann*, though perhaps it is very rare in that country as it is given in a fupplementary feries of plates to his work, and his plates are but a fupplement of the more rare Infects, not figured in Roefel's publications.

PHALÆNA

PHALÆNA AURANTIAGO.

ORANGE MOTH.

LEPIDOPTERA.

PHALÆNA.

SPECIFIC CHARACTER.

Upper wings orange colour with fpots, waves, and ftreaks of brown; feveral minute white fpots along the anterior margin. Body and lower wings cream colour, with a pale wave in the middle of the latter.

———————

This is certainly a non-defcript. *T. Marham, Efq. Sec. L. S.* has defcribed it in his manufcript notes under the fpecific name Aurantiago.

The fpecimen from which the figures in the annexed plate are copied was found on an oak in Richmond Park, in June, 1793. The under-fide as well as upper-fide is fhewn in the plate.

P L A T E

151

PLATE CLI.

FIG. I.

MUSCA BRASSICARIA.

CYLINDRICAL FLY.

DIPTERA.

Wings 2.

GENERIC CHARACTER.

A foft flexible Trunk, with lateral lips at the end.   No Palpi.

SPECIFIC CHARACTER

AND

SYNONYMS.

Thorax greenifh.   Abdomen cylindrical; fecond and third Segment reddifh yellow.

MUSCA BRASSICARIA: antennis fetariis nigra, abdomine cylindrico:
   fegmento fecundo tertioque rufis.   *Fab. Spec. Inf.*
   2. 36. 443.—*Syft. Ent.* 25. *p.* 88.—*Mant. Inf. I.* 2.
   43. 345.—*Ent. Syft.* 4. 327. 63.
Mufca cylindrica: Antennis fetariis pilofa cinereo nigra, abdomine
   cylindrico elongato medio rufo.   *Degeer. Inf.* 6. *n.*
   9. *p.* 30. *tab.* 1. *fig.* 12.
Mouche cylindrique.   Ibid.
Die Kohlfliege.   *Panz. Faun: Inf. Germ.*

C

Тhe

The Mufcæ, if we follow the arrangement of Linnæus, form by far the moft extenfive of any genus (except Lepidoptera) we have at prefent any knowledge of. Fabricius enumerates in his laft * Work no lefs than 202 Species, under the generic title *Mufca* ; independent of thefe we find 122, under the head *Syrphus* ; 22 under *Rhagio*, and 25 under *Stratiomys*, all of which (with fome exceptions) would make by Linnæan arrangement 349 fpecies ; not to notice the Infects of the fame tribe included under his generic appellations, *anthrax*, *bibio*, *nemotelus*, &c.

Mufca Brafficaria is not uncommon in gardens in May and June. Sometimes found on Willows.

Fig. 1. One of the Antennæ magnified.

* *Syft. Ent.*

F I G.

F I G. II.

APIS TUMULORUM.

SMALL, LONG HORNED BEE.

HYMENOPTERA.

Wings four, generally membraneous.   Tail of the female armed with a fting.

GENERIC CHARACTER.

Jaws, with a Trunk bent downwards.   Antennæ elbowed in the middle. Wings plain. Body hairy. Abdomen connected by a pedicle.

SPECIFIC CHARACTER

AND

SYNONYMS.

Antennæ rather longer than the body.   Entirely black, with greyifh hairs.   Jaws yellow.

APIS TUMULORUM.   *Lin. Syft. Nat.* 2. 953. 2. *edit.* 3.—*Fn. Sv.* 1685.

Apis Tumulorum : Antennis filiformibus longitudine corporis nigri, maxillis flavis.   *Fab. Syft. Ent.* 388. 57.—*Spec. Inf* 1. 486. 122.

Eucera Tumulorum, vol. 2. 344. 159.

Sulz. Hift. Inf. tab. 27. *fig.* 14.

This extraordinary Bee is found in Summer, againft banks, when the weather is fine.   Were it not for the remarkable length of the Antennæ, it would fcarcely deferve notice, though it is rather a fcarce Infect.

F I G.

PLATE CLI.

21

FIG. III.

TABANUS PLUVIALIS.

SPECKLED-WING. STINGING FLY.

DIPTERA.

Wings 2.

GENERIC CHARACTER.

Antennæ conic, of four Segments. Trunk flefhy, terminated by two lips. Palpi one on each fide of the Trunk.

SPECIFIC CHARACTER

AND

SYNONYMS.

Eyes green. Thorax brown grey, with feven longitudinal lines. Abdomen grey with marks of black. Wings brown fpeckled with white.

TABANUS PLUVIALIS. *Lin. Syft. Nat.* 16. *p.* 1001. *edit.* 13. *n.* 16. *p.* 2885.—*Fn. Sv. n.* 1887.

Tabanus Pluvialis : Oculis fafciis quaternis undatis, alis fufco punc tatis. *Fab. Syft. Ent. n.* 16. *p.* 790.—*Spec. Inf.* 2. *n.* 23. *p.* 459.—*Mant. Inf.* 2. *n.* 26. *p.* 356.—*Ent. Syft. vol.* 4. *p.* 369. 134. 32.

Tabanus fufcus, alis cinereis, punctis numerofiffimis albis. *Geoff. Inf. T.* 2. *n.* 5. *p.* 461.

Le Taon à ailes brunes piquées de blanc. *Geoff. Inf.*

Die Regenbreme. *Panz. Faun. Inf. Germ.*

Reaum. Inf. 4. *tab.* 18. *fig.* 1.

Harris Inf. angl. tab. 7. *fig.* 8.

C 3 *Scop.*

Scop. carn. n. 1012.
Schrank. Inf. auftr. n. 978.
Schäffer. Icon. Inf. Ratifbon. tab. 85. *fig.* 8. 9.

———————

During all the Summer months we find this tormenting little Infect in great abundance, in the narrow lanes and fkirts of woods, If it fettles on the hands, face, or legs, its fting is very acute, and excites an inflammation and fwelling in the ftung-part, very fimilar to that we experience from the fting of the *Tabanus cæcutiens,* defcribed in Plate 131, of this Work.

Its fting is moft violent about the middle of the day.

F I G.

PLATE CLI. 23

FIG. IV.

MUSCA BOMBYLANS.

DIPTERA.

MUSCA.

SPECIFIC CHARACTER

AND

SYNONYMS.

Antennæ feathered. Black and hairy; extremity of the Abdomen yellow.

MUSCA BOMBYLANS. *Lin. Syst. Nat.* 25. *p.* 983.—*Fn. Sv. n.* 1792.
Syrphus bombylans : Antennis plumatis tomentofus niger, abdomine
poftice rufo.—*Fab. Syst. Ent. n.* 1. *p.* 762.—*Spec.
Inf.* 2. 1. *p.* 421.—*Mantiffa Inf.* 2. 1. *p.* 334.—
Ent. Syst. 4. *p.* 279. 232.
Conops pocopyges. Pod. Muf. græc. n.
Die hummelartige Schwebfliege. *Panz. Faun. Inf. Germ.*
Harris. Inf. angl. tab. 10. *fig.* 3.

This is a common Fly; and is found in woods in May. A figure of one of the Antennæ is given at Fig. 4.

PLATE CLI. 25

FIG. V.

MUSCA TRILINEATA.

TRILINEATED FLY.

DIPTERA.

MUSCA.

SPECIFIC CHARACTER

AND

SYNONYMS.

Yellow green. Three longitudinal black lines on the Thorax. Abdomen marked with black. Two teeth on the scutellum.

MUSCA TRILINEATA : Antennis filatis clavatis, scutello bidentato, corpore viridi, thorace lineis abdomineque faciis nigris. *Lin. Syst. Nat. n.* 6. *p.* 980. *edit.* 13. *n.* 6. *p.* 235.

Stratiomys trilineata : Scutello bidentato, corpore viridi, thorace lineis abdomineque faciis nigris. *Fab. Syst. Ent. n.* 7. *p.* 760.—*Spec. Inf.* 2. 9. *p.* 418.—*Mantiffa. Inf.* 2. 14. *p.* 331.

Stratiomys luteo-virefcens. *Geoff.* Paris. T. 2. n. 7. p. 482.

Stratyomys fafciata. *Fourcroy. Ent.* Paris. 2. 7. p. 468.

Grüne Waffenfliege. Dreygeftreifte Waffenfliege.

La Mouche-armée jaune à bandes noires. *Panz. Faun. Inf. Germ.*

A very curious and fcarce fpecies. It was found among fome elder leaves which were gathered in Batterfea Meadows, early in June, 1795.

When

When this Infect is alive the yellow colour of the body is exceedingly bright, and partakes somewhat of a metallic and green hue in several parts, but this brilliant appearance gradually fades after death.

The line denotes the natural fize, it being neceffary to give a magnified figure of fuch a minute Infect.

P L A T E

152

PLATE CLII.

PHALÆNA AESCULI.

WOOD LEOPARD MOTH.

LEPIDOPTERA.

GENERIC CHARACTER.

Antennæ taper from the bafe. Wings in general deflexed when at reft. Fly by night.

SPECIFIC CHARACTER
AND
SYNONYMS.

Wings white, with many dark blue round fpots. Six fpots on the Thorax.

PHALÆNA AESCULI elinguis lævis nivea, antennis thorace brevio-
ribus, alis punctis numerofis cœruleo nigris, thorace
fenis. *Lin. Syft. Nat.* 2. 833. 83.—*Fn. Sv.* 1150.
Bombyx Aefculi, Mant. Inf. 2. 116. 85.
Hepialus Aefculi. Fab. Spec. Inf. 2. 208. 146. 4.
Coffus Aefculi. *Wien. Verzeichn. tab. tit. præf. Acta Soc. Berol. phys.*
3. *tab.* 1. *fig.* 1. 2.
Pod. Inf. 88. 16.
Wood Leopard Moth. *Harris Inf. angl.*

― ― ― ― ―

It is to a very fingular and trivial circumftance we are indebted for the fpecimens of both the male and female of this rare fpecies. They were obferved together on the bark of an elm tree in the Mall in St. James's Park, by fome ignorant perfons, who being terrified at their extraordinary appearance, attempted to deftroy them, but a
gentleman

gentleman who happened to pafs by at the fame inftant, having either more curiofity or lefs apprehenfion of danger from touching them, took them up, and preferved them. We conclude they could have but juft before come out of their chryfalides, the female being in a moft perfect ftate, and the male equally fine, except that it had loft one of its upper wings.

We muft claim the indulgence of the more fcientific part of our readers for the minutenefs with which we have detailed fuch trifling circumftances; it can indeed afford very little amufement to them, but, it may ferve to remind many who are not in the habit of collecting Infects, that their occafional endeavours would be likely to extend the Science of Entomology; for it often happens that the moft affiduous Naturalifts are indebted to fuch perfons for the rareft fpecimens their cabinets can boaft.

The Moths were found late in June. On examining the crevices of fome of the trees near the fpot, we found a quantity of the eggs; they were rather of an oval form, and linked together like a chain, as fhewn in the Plate; and having carefully preferved them in a branch of a plumb-tree * under the bark, we had the fatisfaction to fee fome young Caterpillars produced in a few weeks. But either owing to the want of proper food or good management they all died foon after except two or three, and thefe never arrived at their full fize. The Caterpillar from which the Figure in the annexed is copied, was found under the bark of one of the elm-trees in St. James's Park, but being difturbed, it never became a Pupa. The Caterpillar makes a cafe, of the duft of the wood which it gnaws, and cements together, and in this it lies concealed beneath the bark. The head of the Caterpillar is hard, and the firft ring is furnifhed with a ftrong horny fubftance.

Harris, about twenty years ago, was fo fortunate as to breed this Moth from the Caterpillar, and we are not acquainted with any

---

* I frequently find, when the Plant of an Infect is unknown, that they will live on the Plumb-tree, when they refufe other food.

fimilar

PLATE CLII.   29

fimilar inftance fince that time. In the Plates of Roefel, vol. 4, a Figure of the Caterpillar is given, but without either Pupa or Moth, fo that were it not for the reference and authority of Linnæus, and fince his time, of Fabricius, it would fcarcely be known to what Infeƈt it belonged. The eggs we have not found either figured or defcribed, though they are fo very fingularly united together, and would certainly have been noticed by the ingenious Roefel if he had met with them.

The Antennæ of the female are fetaceous, or like a briftle, but that part of the male is both fingular and beautiful; it is elegantly feathered next the bafe, and terminates in a briftle, like the female.

PLATE

153

P L A T E   CLIII.

F I G. I.

PHALÆNA EUPHORBIATA.

SMALLEST QUAKER MOTH.

LEPIDOPTERA.

GENERIC CHARACTER.

Antennæ taper from the bafe. Wings in general deflexed when at reft. Fly by night.

GEOMETRA.

SPECIFIC CHARACTER

AND

S Y N O N Y M S.

Entirely brownifh grey without fpots.

PHALÆNA EUPHORBIATA : feticornis alis fufco cinereis immacu-
latis. *Fab. Mantf.* 2. *p.* 209. *n.* 202. *Ent. Syft.*
T. 3. *p.* 2. 197. 246.
DE VILLERS Ent. Lin. *T.* 4. p. 509. *De l'Euphorbe.*
Hubners Beitr. 1. B. 2. Th. p. 15. *Ph. G. unicolorata.* Tab. 3. fig.
L. 2. B. 4. Th. p. 112.
Langs Berz. p. 189. n. 1361. 62. Ph. *G. unicolorata.* Der Klein-
grave Nachtfalter.
Berlin. Magaz. 4. Th. p. 524. n. 44. *Ph. fafcata.* Der Sperling.
Der Wolfsmilchfpanner. *Klem. Inf. Suppl. T.* 2. Tab. 24. fig. 1.
Wien. Verz. 116. 9.
Hybn. Beytr. 2. tab. 3. fig. L.

I   This

This is not an uncommon Moth in fome places, yet we find no figure of it in any work on Britiſh Inſects. In the work of Klemann, quoted in the Synonyms, a figure of it is given without the Larva ; from this we may ſafely infer it is ſeldom found in that ſtate, or that indefatigable writer would certainly have added it to his Plate.

It is ſuppoſed to feed on ſome plant of the Euphorbia genus, and hence the ſpecific names fuſcata and unicolorata have been abandoned.

The Moth was found late in May.

F I G.

F I G. II, III, IV.

PHALÆNA UDDMANNIANA.

CHESNUT SPOT MOTH.

LEPIDOPTERA.

PHALÆNA.

Tortrix. Lin.
Pyralis. Fab.

SPECIFIC CHARACTER

AND

SYNONYMS.

Wings greyish brown. An angular shaped chesnut coloured spot on the posterior margin of the first pair.

TORTRIX UDDMANNIANA: alis cinereis: macula brunnea communi transfversa. *Lin. Syst. Nat.* 2. 880. 320.— *Fn. Sv.* 1332.

Pyralis Uddmanniana. Fab. Spec. Inf. 2. *p.* 279. *n.* 22.—*Mant. Inf.* 2. *p.* 228. *n.* 35.

Wiener Verz. p. 130. Fam. D. grave Blattwickler (Ph. Tortrices cinereæ) &c. *l'Uddmann.* de VILLERS ent.

Der himbeer unkler. *Kleman. Inf.* Suppl. Tab. 24.

De PRUNNER larv. d'Eur. p. 35. *Tortrix Uddmanniana.*

———————————

This is much rarer than the preceding species, and is also a far more beautiful Insect. We have never found it except about the

D

hazel

hazel nut trees in Coombe Wood, Surry, though it may, no doubt, be met with wherever thefe trees are found in abundance. Is found in Germany.

The Caterpillar changed to Chryfalis in May. Moth appeared in July.

FIG.

PLATE CLIII.   35

FIG. V.

PHALÆNA CARNELLA.

ROSE COLOURED VANEAR.

LEPIDOPTERA.

PHALÆNA.

Tinea.

SPECIFIC CHARACTER

AND

SYNONYMS.

Upper Wings rose colour, anterior margin whitish, posterior yellowish. Lower Wings pale.

TINEA CARNELLA : alis anticis flavis: lateribus sanguineis. *Lin.*
     Syst. Nat. 2. 887. 353.—*Fab. Spec. Inf.* 2. 293. 21.
     Ent. Syst. 3. Pars. 2. 296. 41.
     Wien. verz. 138. 13.
     Schœff. Icon. tab. 147. 2. 3.
     Sulz. Hift. Inf. tab. 23. *fig.* 12.
Purple Vanear ? Harris. Inf.

———————

The Larva of this rare and elegant Insect is wholly unknown to Collectors of British Insects. The Moths were formerly taken at the Chalk-pits, near Charton, in Kent, but either the brood has been destroyed, or the seasons so unfavourable, that few, if any, have been seen for several years. The Moth comes forth in May, and, like other species of the same tribe, fly very low, and always settle on the blades of grass, with their Wings folded, so that Collectors can readily distinguish them from other Moths.

PLATE

154

PLATE CLIV.

FIG. I, II, III.

SCARABÆUS NOBILIS.

SCARCE GREEN CHAFFER.

COLEOPTERA.

Wings two, covered by two fhells, divided by a longitudinal future.

GENERIC CHARACTER.

Antennæ clavated, extremities fiffile *.   Five joints in each foot.

SPECIFIC CHARACTER

AND

SYNONYMS.

Shining green ; fhells, full of wrinkles.   Thorax not projecting.

SCARABÆUS NOBILIS : fcutellatus muticus auratus, abdomine poftice albo punctato. *Linn. Syft. Nat.* 2. 558. 81.— *Fn. Sv.* 401.

Cetonia nobilis : aurata, abdomine poftice albo punctato, elytris rugofis. *Fabr. Syft. Ent.* 43. 5.—*Spec. Inf.* 1. 6. *p.* 51.

Scarabæus viridis nitens, thorace, infra æquali, non prominente. *Geoff. Inf.* 1. 73. 6.

Scarabæus auratus fecundus. *Roef. Inf.* 2. *Scarab.* 1. *tab.* 3. *fig.* 1, 2, 3, 4, 5.

Scarabæus viridulus fcutellatus aureo viridis nitidus, elytris rugofis abdomine poftice albedine maculato, pectore mutico. *Degeer. Inf.* 4. 297. 26.

---

* Divided into laminæ, or parts.

E   This

This fpecies is not much unlike the Scarabæus Auratus (large green Beetle, or Rofe Chaffer) but is far more fcarce. The larva lives entirely under the furface of the ground, and feeds on fmaller Infeets. The Jaws are very ftrong, but in other refpeets it appears unable to defend itfelf if attacked. It is very fluggifh, and always lies with its body coiled round. The cafe in which it remains in the pupa ftate is very ftrong, and confifts of fmall bits of wood, pebbles, earth, &c. cemented and faftened together, by a flight filky web. It continues during the Winter in this cafe, and in May the Beetle comes forth.

Fig. 1. The larva.   Fig. 2. Pupa.   Fig. 3. Perfeet Infeet.

SCARABÆUS

PLATE CLIV.   39

FIG. IV.

SCARABÆUS LUNARIS,

LUNATED BEETLE.

COLEOPTERA.

SCARABÆUS.

SPECIFIC CHARACTER

AND

SYNONYMS.

Entirely black. On the head a lunated helmet and an erect horn. Thorax with three horns; the center one obtuse and divided by a longitudinal furrow. Eight furrows down each shell.

SCARABÆUS LUNARIS: exscutellatus, thorace tricorni, intermedio obtuso bifido, capitis cornu erecto. clypeo emarginato. *Linn. Syst. Nat.* 2. 543. 10.— *Fn. Sv.* 379.
         Fab. Spec. Inf. 1. 24. 108.
Copris capitis clypeo lunulato, margine elevato, corniculo denticulato. *Geoff. Inf.* 1. 88. 1.
Scarabæus ovinus tertius f. capite operculato. *Raj. Inf.* 103.
Scarabæus naficornis medius. *Frifch. Inf.* 4. 25. *tab.* 7.
         Pet. Gazoph. tab. 138. *fig.* 4.
         Schœff. Icon. tab. 63. *fig.* 2. 3. ♂ . ♀ .
         Bergftræff. Nomencl. 1. 5, 9. *tab.* 1. *fig.* 9. *et tab.* 4. *fig.* 7.

This is by no means a common Beetle. It is found generally amongst the loose sand on heaths, the dung of animals, or carrion. The female is nearly as large as the male, and has not the erect horn on the head.

E 2   PLATE

PLATE CLV.

SPHINX STELLATARUM.

HUMMING-BIRD HAWK-MOTH.

LEPIDOPTERA.

GENERIC CHARACTER.

Antennæ thickeſt in the middle. Wings, when at reſt, deflexed. Fly morning and evening only.

SPECIFIC CHARACTER

AND

SYNONYMS.

Abdomen thick, brown, and hairy ; tufted at the extremity. Firſt Wings greyiſh brown, with waves of black acroſs. Second Wings orange colour.

SPHINX STELLATARUM. *Linn. Syſt. Nat.* 2. 803. 27.—*Fn. Sv.* 1094.
Seſia Stellatarum : abdomine barbato, lateribus albo nigroque variis,
            alis poſticis ferrugineis. *Fab. Syſt. Ent.* 548. 3.
            Fab. Spec. Inſ. 2. 154. 6.
Papilio velociſſima, alis albis brevibus, corpore craſſo inter volitandum
            ſtridorem edens. *Raj. Inſ.* 133. 1.
      Roeſ. Inſ. 1. *papilionum Noĉturnorum.* *Tab.* 8.
      Bradl. Nat. tab. 26. *fig.* 1. A.
      Reaum. Inſ. 1. *tab.* 12. *fig.* 5. 6.
          E 3   *Merian.*

PLATE CLV.

Merian. Europ. 2. 33. tab. 29.
Schœff. Elem. tab. 116. fig. 3.
———— Icon. tab. 16. fig. 1.
Le Colibri. *Harris. Aurel. pl. 24.*

There are two forts of Caterpillars belonging to this fpecies. They are alike in fize and form, but are very different in colour. One fort is green, the other purplifh red, varying much in different fpecimens, being fometimes almoft brown. Both forts are fpotted with minute white fpecks, which are difpofed in regular order over every part, except the belly.

Every Caterpillar is alfo furnifhed with a pofterior horn, which is blue from the bafe for more than half its length : the tip is bright orange colour.

The Chryfalis, which is of a pale yellowifh-brown at firft, changes to a more dufky colour before the Sphinx comes forth.

The Caterpillars feed on feveral kinds of plants, but feem chiefly to prefer thofe of the *Galium* genus, particularly, the White * or Yellow † Lady's Bedftraw, and Cleaves, or Goofegrafs ‡. They go into the ground about the latter end of Auguft, and remain there in chryfalis till April, or May at the fartheft.

It is rather a fcarce Infect : fometimes vifits gardens in the winged ftate ; and extracts the fweeteft juices of the flowers, by darting its long probofcis, or trunk into them ; it is from this peculiarity, and its hovering over the flowers at the fame time, like the Humming Birds when they feed, that it has received its Englifh appellation.

---

* *Galium Paluftre.*
† ———— *Verum.*
‡ ———— *Aperine.*

This

This Insect is found in most parts of Europe, but it appears is more frequent in Northern Countries. A near variety of it is found in Botany-Bay ; and we have specimens of it from North America.

Sphinx Belis of Linnæus and Cramer, is described amongst the Synonyms given by Fabricius, as a variety of *Sphinx Stellatarum,* and *Sphinx Ciculus* of Cramer scarcely differs from our Insect.

E 4

PLATE

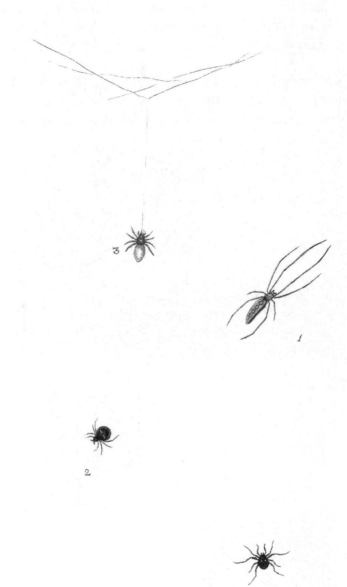

PLATE CLVI.

FIG. I.

ARANEA EXTENSA.

APTERA.

No Wings.

ARANEA.

SPECIFIC CHARACTER

AND

SYNONYMS.

Abdomen long, greenifh, and filvery.   Legs very long.

ARANEA EXTENSA : abdomine longo argenteo virefcente, pedibus longitudinaliter extenfis.   *Linn. Syft. Nat.* 2. 1033. 22. *Fn. Sv.* 2011.

Aranea retiaria abdomine elongato grifeo fufco, pedibus longitudinalibus extenfis.
> *Degeer. Inf.* 236. 1.
> *Geoff. Inf.* 2. 642. 3.
> *Lift. Aran. fig.* 3.
> *Raj. Inf.* 19. 3.

This fpecies is particularly diftinguifhed by the length and pofition of its legs.   It runs very faft.   Our fpecimen was taken on an oak, and we do not think it is a ground Spider.

Found in Darent wood, Dartford, in Auguft.

FIG.

PLATE CLVI.

F I G. II.

ARANEA GLOBOSA.

GLOBULAR SPIDER.

APTERA.

GENERIC CHARACTER.

Legs eight. Eyes eight.

SPECIFIC CHARACTER

AND

SYNONYMS.

Black. Abdomen globular, fides crimfon.

ARANEA GLOBOSA: nigra abdominis lateribus fanguineis. *Fab. Ent. Syft.* 2. 411. 15.

We have met with this beautiful Spider feveral times in Caenwood. It was commonly feen on the young oaks. One being confined in a box fpun a fmall web, of a very flight texture. Found in May and June.

F I G.

F I·G. III.

ARANEA CINEREA.

APTERA.

ARANEA.

SPECIFIC CHARACTER

AND

SYNONYMS.

Abdomen ash colour, or grey.   Thorax and feet yellow-brown.

ARANEA CINEREA : abdomine cinerascente.   Thorace pedibusque
testaceis. *Panzer.*
Die afchgrave Spinne. *Panz. Inf. German.*
Aranea Cicurea, pallide rubra abdomine ovato cinereo. *Fab. Ent.*
Syft. 2. 410. 12?

―――――――――――――

A common Spider in woods.   Found in May and June.

F I G.

PLATE CLVI.

FIG. IV.

PHALANGIUM BIMACULATUM.

MINUTE BLACK SPIDER, WITH TWO WHITE SPOTS.

APTERA.

No wings.

GENERIC CHARACTER.

Legs eight, eyes two.   Abdomen rounded.

SPECIFIC CHARACTER

AND

SYNONYMS.

Very minute.   Entirely black, except two white ſpots on the Thorax.

PHALANGIUM BIMACULATUM : abdomine atro: maculis duabus
albis.   *Fab. Ent. Syſt. v. 3. n. 8. p. 431.*
Die zwey fleckigte Afterſpinne.   *Panz. Faun. Inſ. Germ.*

This is a very minute Inſect ; the figure is more than twice the natural ſize.   It was found amongſt a great variety of other ſpiders, in Darent wood, Dartford, about the middle of Auguſt.

157

P L A T E CLVII.

PHALÆNA BERGMANNIANA,

LEPIDOPTERA.

GENERIC CHARACTER.

Antennæ taper from the bafe. Wings in general deflexed when at reft. Fly by night.

Tortrix *Linn.* Pyralis *Fab.*

SPECIFIC CHARACTER

AND

SYNONYMS.

Firft wings yellow, varied with orange colour. Four brown marks acrofs each wing, with fpots and ftreaks of filver down them. Inferior wings grey.

PHALÆNA BERGMANNIANA. *Linn. Syft. Nat.* 2. 878. 307. *Fn. Sv.* 1314.

PHALÆNA BERGMANNIANA: alis anticis luteis flavo punctatis, fafciis quatuor argenteis, tertia bifida. *Fab. Syft. Ent.* 652. 43. *Spec. In.* 2. 285. 59.

Phalena antennis filiformibus, alis luteis nitidis, ftrigis quatuor argenteis.

Phaléne à antennes filiformes à trompe à ailes larges d'un jaune orange luifant avec quatre rayes tranfverfes d'un brun argenté. *Phaléne chappe jaune à rayes argentées. Degeer Inf.* 2. *p.* 1. *p.* 469. *n.* 4.—*Inf.* 2. 1. 346. 4.

Phal. Pallium aurantium. fpirilinguis, antennis filiformibus. alis rhombeis aurantiis nitidis ftrigis 4 fufco argenteis. RETZ. *Degeer, p.* 52. *n.* 147.

Phal.

Phal. Bergmanniana.   Alæ anticæ flavæ nodulis binis, fafciifque (4) argenteis margine fufco-ferrugineo.   Scopoli *ent Carn. p.* 232. *n.* 584. *fig.* 584.

Tortr. eur. *Bergmanniana. Jungs alphab. Berf.* 2. *Th. p.* 75.

Tortrix Bergmanniana. la Bergmann. *de* Villers *ent. Lin. T.* 2. *p.* 396. *n.* 671.

Der Bergnannfche unkler.   *Kleeman Inf. Nr.* 45. 1794.

Metallifche Blattwicktler (*Phal. Tortrices Metallicæ*) n. 5. Tortrix. Bergmanniana *Wiener. Verz. p.* 126. *Fam. B.*

Bergmannfwicktler.   *Brahms Hanbd.* 2. *Th.* 1. *Ubth. p.* 237. *n.* 132

Der Bergmannifche Nachtfalter.   *Langs Verz. p.* 203. *n.* 1379, &c.

───────────

Linnæus gave this little Moth the fpecific name *Bergmanniana,* in honour of Prof. Bergmann, a naturalift of diftinguifhed eminence. It is a very pretty Infect; but, when magnified, its appearance is truly fuperb, the ground colour which is bright yellow, fhewing the orange markings to great advantage, and the metallic fplendour of the burnifhed filver appearing like raifed work above the ftripes or bands of dark brown that crofs the upper wings.

We have found this Moth at Highgate. The Caterpillars are yellow, with a ftreak of green down the back; but the green difappears before the laft fkin, in which they are of a pale yellow, without any marks whatever. They feed on white thorn.

Fig. 1, 2. The Caterpillars. Fig. 3. Chryfalis. Fig. 4. The fame magnified. Fig. 5. Moth. Natural fize. Fig. 6. The fame magnified.

F I G.

F I G. VII.

P H A L Æ N A S Q U A M A N A.

GREEN TUFTED, OR BUTTON MOTH.

LEPIDOPTERA.

PHALÆNA.

Tortrix *Lin.*

SPECIFIC CHARACTER

AND

SYNONYMS.

Upper wings green, tufted all over. Inferior wings pale brown.

PHAL. PYRALIS SQUAMANA : alis virefcentibus fcabris. *Fab. Syft.*
Ent. 651. 36. *Spec. Inf.* 2. 284. 50.

———

This is exceedingly rare. The upper wings are very curious, being entirely covered with tufts of feathers, of various fizes, fome brownifh, others inclining to white, but moft of them are green, which is the ground colour of the wings. Of its Larva we are entirely ignorant ; nor can we derive any affiftance in that refpect from entomological writers, as *Fabricius* only has defcribed the Moth. He fays it is a native of England, and preferved in the cabinet of Mr. *Monfon.*

Taken in June.

P L A T E

158

PLATE CLVIII.

PHALÆNA VERSICOLORA.

GLORY OF KENT MOTH.

LEPIDOPTERA.

GENERIC CHARACTER.

Antennæ taper from the bafe, Wings in general deflexed when at reft.   Fly by night.

Bombyx.

SPECIFIC CHARACTER

AND

SYNONYMS.

Antennæ feathered.   Male, firft wings red brown, with tranfverfe waves, black and white lines, and three white fpots at the extreme angle.   Second wings orange.   Female larger, and colours paler throughout.

PHALÆNA VERSICOLORA: *Lin. Syft. Nat.* 2. 817. 31. *Fn. Sv.*
        IIII.

BOMBYX VERSICOLORA: alis reverfis grifeis nigro-albis thorace
        antice albo.   *Fab. Syft. Ent.* 565. 34.—*Spec.*
        Inf. 2. *n.* 50. *p.* 178.—*Mant. Inf. T.* 2. *n.* 58.
        p. 113.

Phalæna alis lineis albis et nigris undatis.   *Gadd. Satag.* 82.
        Roef. Inf. 3. *tab.* 39. *fig.* 3.
        Sulzer Hift. Inf. tab. 21. *fig.* 4.
        Fueft. Magaz. 2 *tab.* 1. *fig.* 4.

Der Buntflügel.   Der Hagebuchenfpinner.
Das Männchen.   *La Verficolore.*   (*Male.*)
Das Weibchen.   (*Female.*)   *Panz. Faun. Inf. German.*

F.   This

This extremely rare Infect is always confidered as a Britifh fpecies, and is ufually found in the cabinet of the Englifh entomologift ; yet thofe are German Infects generally, for we know only of one fpecimen which is clearly afcertained to have been found in this country. The fpecimen alluded to is in the collection of Mr. Francillon, jeweller, in Newcaftle-ftreet, in the Strand : it is a female, and was found by that gentleman's brother in his garden at Carfhalton.

Whether Harris ever met with this Infect we cannot pretend to determine ; he fays it appears in the winged ftate in April *. We cannot hefitate to fuppofe, that this Moth has been found in England feveral times, particularly in *Kent* ; but none of thefe remain at this period in the collections of the curious.

The Male differs much from the Female : both fexes are fhewn in the annexed plate.   Fig. I. Male.   Fig. II. Female.

Fuefly, in a German publication, has given the only figure we are acquainted with of the Caterpillar of this Moth ; and *Fabricius* has copied his defcription from the coloured engraving. It is green, with oblique lines of yellowifh brown, and large fpots of golden yellow.

---

* *Vide Aurelian's* Companion.

P L A T E

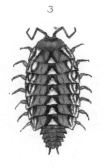

159

3

2

1

4

PLATE CLIX.

FIG. I, II, III, IV.

ONISCUS AQUATICUS.

APTERA.

No Wings.

GENERIC CHARACTER.

Legs fourteen. Antennæ taper. Body oval.

SPECIFIC CHARACTER

AND

SYNONYMS.

Afh colour. Antennæ of four joints. At the end of the tail two bifid appendices.

ONISCUS AQUATICUS: cauda rotundata, ftylis bifurcis, antennis quaternis. *Syft. Ent.* 297. 6.—*Spec. Inf.* 1. 376. 6.

Onifcus aquaticus lanceolatus, cauda rotundata, ftylis bifurcis. **Linn.** *Syft. Nat.* 2. 1061. 11.—*Fn. Sv.* 2061.

Squilla Afellus aquatica, cauda rotundata, ftylis binis bifurcis. **Degeer.** *Inf.* 7. 496. 1. *tab.* 31. *fig.* 1.

Afellus aquaticus *Gefneri.* *Raj. Inf* 43. 1.

Sulz. Hift. Inf. tab. 30. *fig.* 12.
Frifch. Inf. 10. *tab.* 5.
Schæff. Elem. tab. 22.

This fpecies is lefs frequent than Onifcus Afellus, (Common Woodloufe). It lives in clear waters, moft part of the fummer. It

G

fcarcely

fcareely exceeds one half of the length of O. Afellus in England, yet if we may form an opinion of the German fpecimens from thofe figured by *Sulz*, they are larger than with us.

The Onifcus Agilis of *Perfoon*, figured in Panzer's Work *, correfponds perfectly with ours in fize ; and the minute markings on the fhells, if carefully examined with a glafs, will be found nearly alike.   The antennæ of the figure in Sulz feems rather contrary to the fpecific character of the infect ; and that of Panzer's, though of another fpecies, more refemble thofe of our fpecimen.

Of the *Onifcus Afellus* we find different coloured fpecimens, fome are almoft white with grey marks, others are nearly deep black ; we find alfo, *Onifcus Aquaticus* liable to variations, though not fo much as the former infect in fome the light ground colour is very diftinct, in others rather confufed.   Some are deeper coloured ; and again, many, when firft taken, have a fine glowing, olive brown appearance throughout, though lefs vivid than that of Onifcus Agilis before noticed.

Fig. 1. 2. Natural fize.   Fig. 3. Magnified.   Fig. 4. Antennæ.

---

* *Faun. Inf. Germ.*

PLATE

160

PLATE CLX.

PHALÆNA PUDIBUNDA.

PALE TUSSOCK MOTH.

LEPIDOPTERA.

GENERIC CHARACTER.

Antennæ taper from the bafe. Wings in general deflexed when at reft. Fly by night.

Bombyx.

SPECIFIC CHARACTER

AND

SYNONYMS.

Wings light, greyifh : three tranfverfe waves acrofs each upper wing.

PHALÆNA PUDIBUNDA : alis deflexis cinereis, ftrigis tribus undatis fufcis. *Lin. Syft. Nat.* 2. 824. 44.

Fn. Sv. 1118.

Fab. Spec. Inf. 2. 183. 68.

Ent. Syft. Tom. 3. *p.* 1. *p.* 438. 97.

Phalæna pectinicornis, elinguis, alis deflexis cinereo undulatis,
fafciis tranfverfis obfcurioribus, capite
inter pedes porrectos. *Geof. Inf.* 2.
113. 15.

Phalæna cinerea, alis oblongis, exterioribus quatuor lineis nigrican-
tibus tranfverfis, diftinctis. *Raj. Inf.*
185. 7.

Roef. Inf. 1. *phal.* 2. *tab.* 38,

Ammir. tab. 18.

Goed. Inf. 3. *tab.* 5.

Merian Europ. 1. *tab.* 47.

Degeer Inf. 1. *tab.* 16. *fig.* 11. 12.

The

The light Tuffock Moth is found late in September, or during the month of October. The Caterpillar is both beautiful and fingular: it feeds on the oak, on which it is met with, from the latter end of July till the middle of September, at which time it is of its full fize, and becomes a pupa; it fpins a web between the leaves, and remains in the chryfalis about thirty days. The eggs are of a pale brownifh colour, fig. 1.

P L A T E

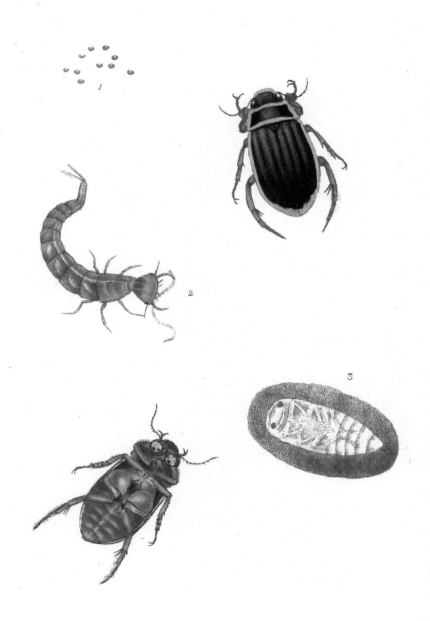

161

PLATE CLXI.

DYTISCUS MARGINALIS.

LARGE BOAT BEETLE.

COLEOPTERA.

Wings two, covered by two fhells, divided by a longitudinal future.

GENERIC CHARACTER.

Antennæ taper, or clavato-perfoliated. Feet villous and broad.

SPECIFIC CHARACTER
AND
SYNONYMS.

Black; exterior margin of the thorax and fhells yellow. Eyes large, round, black.

DYTISCUS MARGINALIS: niger thoracis marginibus omnibus elytrorumque exteriori flavis.
> *Lin. Syft. Nat.* 2. 665. 7.
> *Fn. Sv.* 769.
> *Fab. Spec. Inf.* 1. 291. 3.
> *Ent. Syft. Tom.* 1. 187. 3.

Dytifcus nigro fufcus nitidus, thorace undique elytrorumque margine flavo. *Degeer. Inf.* 4. 391. 2. *tab.* 16.
> *fig.* 2.

Hydrocantharis noftras. *Raj. Inf.* 93. 1.
> *Mouff. Inf.* 164.
> *Lift. Mut. tab.* 5. *fig.* 42.
> *Sulz. Hift. Inf. tab.* 6. *fig.* 42.
> *Roef. Inf.* 2. *Aquat.* 1. *tab.* 1.
> *Schœff. Icon. tab.* 8. *fig.* 7.

β. *Dytifcus femiftriatus* fufcus, elytris fulcis dimidiatis decem.
> *Lin. Syft. Nat.* 2. 665. 8.—*Fn. Sv.* 772.

The

PLATE CLXI.

62

The transformation of any infect from one ftate to another is both curious and entertaining to an enlightened obferver; yet there are a few fpecies whofe manners are fo peculiar, and their changes fo aftonifhing, that they feem to demand more than ordinary attention; and of this defcription we confider the fubject of the annexed plate. If we fpeak of it as to its manners collectively, one peculiarity implies a contradiction of the other, for it is an aquatic, a terreftrial, and an aerial creature. Few infects that inhabit the water, in the perfect ftate ever quit it; and the generality of thofe whofe larvæ live in that element could exift for a few minutes only in it, after they become winged infects; this is particularly noticed of the *Libellulæ*, *Phryganeæ*, *Ephemeræ*, *Tipulæ*, and an immenfe croud of other infects that are bred in the water; but it appears this infect in the larva ftate can leave the water without injury, and in the laft ftate, though a winged creature, it lives for the moft part in the water, and quits it only in the evenings; or when the pool dries up, it ufes its wings in fearch of another.

In the larva ftate it is not lefs remarkable for its favage difpofition, than its formidable appearance. The whole body is covered with a hard fhell, or coat of mail, and the head is armed with two long, femi-circular, fharp-pointed forceps. It is very alert in the water, and when it takes its prey, which confifts of fmaller aquatic infects, it plunges thefe weapons into them, and through a minute aperture, at the extremity, it extracts all their juices. When the time arrives in which it is to become a pupa, it leaves the water and forms a cavity juft below the furface of the earth of an oval form: how long it remains in this cavity in the pupa ftate is uncertain. The beetle comes forth in May.

Much doubt has arofe refpecting the female of this fpecies; Linnæus, in the Syftema Naturæ, defcribed the fuppofed female as β *Dyfticus Semiftriatus*. Fabricius, in the Species Infectorum, adds a long lift of fynonyms from different entomological writers, feveral of whom had figured or defcribed it as a diftinct fpecies before the time of Linnæus, and fome fubfequent authors have held the fame opinion; but in the laft work, *Entomologia Syftema,*. Fabricius confiders it to be

the

PLATE CLXI. 63

the female, and includes only a few of his former references. Upon the firft view of thefe opinions the point feems undetermined; and though we partly affent to the opinion of the laft writer, we muft endeavour to be entirely fatisfied, before we give a figure of Dytifcus Semiftriatus.

The upper fide of this infect is generally defcribed black; this is not the colour in living fpecimens: it is of a fine gloffy black-green, and the marginal colour brighter than in thofe that have been dead fome time. The greenifh hue on the back feldom entirely difappears.

The fore feet of this beetle have an appendage of a very fingular ftructure; it is nearly round, flat beneath, and has in the middle two remarkable circular cavities, with many others more minute: it is fuppofed, that through minute apertures in thefe cavities it can emit a kind of oily fluid; or that, by their affiftance, it can collect air bubbles, to raife itfelf from the deep parts of the water to the furface, in an inftant. The larva of the Mufca Chamælion, which lives in the water, collects the air in a bubble within the rays of its tail, and thereby raifes itfelf to the furface in like manner.

Fig. 1. The eggs. Fig. 2. The larva. Fig. 3. The pupa.

162

2

1

PLATE CLXII.

FIG. I, II.

LEPISMA POLYPODA.

APTERA.

No Wings.

GENERIC CHARACTER.

Legs fix, broad and fcaly at their origin. Palpi two, moveable. Antennæ filiform. Tails three. Body fcaly.

SPECIFIC CHARACTER

AND

SYNONYMS.

Grey, brown, black intermixed; a very high protuberance on the back. Three tails.

LEPISMA POLYPODA: faltatoria, cauda triplici, abdominis feg‑
mentis fubtus utrinque villofis. *Fab. Spec.*
Inf. 1. 380. 2.
Lepifma polypoda fcutata, cauda triplici. *Lin. Syft. Nat.* 2. 1012. 2.
Forticina teres faltatrix. *Geoff. Inf.* 2. 614. 2.
Lepifma fquamofa faltatoria, fetis caudæ tribus intermedia majore.
Strœm. Act. Hafn. 9. 575. *tab.* 2.

Fig. 1. The natural fize. Fig. 2. Magnified.

This is a very rare and curious fpecies; it was found amongft fome loofe ftones, in a damp fituation, July, 1796.

PLATE

163

PLATE CLXIII.

PHALÆNA DISPAR.

GIPSEY MOTH.

LEPIDOPTERA.

GENERIC CHARACTER.

Antennæ taper from the bafe.   Wings in general deflexed when at reft.   Fly by night.

Bombyx.

SPECIFIC CHARACTER

AND

SYNONYMS.

Female, yellowifh white with dark tranfverfe zigzac lines acrofs the upper wings.   Male, fmaller, dark brown, with lines and waves of black.

PHALÆNA DISPAR: alis deflexis mafculis grifeo fufcoque nebulofis, fœmineis albidis: lituris nigris.

> *Lin. Syft. Nat.* 2. 821. 44.
> *Fab. Spec. Inf.* 2 182. 66.
> —*Ent. Syft.* 3. pars. 1. 437. 94.
> *Roef. Inf.* 1 phal. 2. tab. 3.
> *Reaum. Inf.* 2. tab. 1. fig. 11. 14.
> *Merian. Europ.* 1. tab. 18.
> *Frifch. Inf.* 1. 14. tab. 3.
> *Schæff. Icon. tab.* 28. fig. 3—6.
> *Geoffr. Inf.* 2. 112. 14.

That

That the Phalæna Difpar was not uncommonly fcarce about fifteen years ago, is evident from this circumftance, few collections of Britifh infects, that were in the hands of eminent collectors, are without an Englifh fpecimen, which was procured about that time; and Harris, in 1775, as well as fome other writers about the fame period, fpeak confidently of its being found in this country. Berkenhout, in his Synopfis, fays, it is "*frequent* about Ealing, in Middlefex." But this we can, on the beft authority, difpute; it never was frequent in that place, though it has feveral times been met with, by collectors of infects; a parcel of eggs being obtained from them, and hatching, many caterpillars were procured; and thefe being carefully attended, feveral moths were alfo produced. This is not a very extraordinary circumftance, as many of the rareft infects may become common, when the eggs, or a brood of caterpillars, can be difcovered.

We are willing to acknowledge, that we have not been more fortunate in our refearches for the caterpillar or moth of this fpecies, than any others engaged in the fcience of entomology; but we have procured from Germany a collection, containing many valuable rarities that have been found in this country at different times; amongft thefe we have moft perfect and finely preferved fpecimens of *Phalæna Difpar*, in its feveral ftates, and thefe perfectly agree with thofe formerly collected in England. Our Plate contains only one figure of the caterpillar, and that is of the female. The male differs only in being fmaller, and in the fize of the head, which is lefs in proportion than that of the female.

In this inftance, we truft, any apology will be unneceffary, though the original fpecimens were not found in this country: it muft be an advantage to the work to contain figures of the rareft infects; and fhould any of our readers be fo fortunate as to find the caterpillar, they will be able to determine the fpecies, and the proper food to rear it on; or, if the brood be extinct, the plate will be more interefting, as there cannot remain a doubt of its having been indigenous in England.

In

PLATE CLXIII. 69

In foreign countries it is very injurious to gardens, and fruit-trees in particular. *Berkenhout* fays, it feeds on " Oak, Afh, Apple-trees, &c." but we are rather inclined to doubt his information, except as to the latter, though he is partly fanctioned by *Linnæus*. *Geoffroy* fays, it feeds on the Elm.

For the time of its appearance we are indebted to *Harris*; he fays the caterpillar changed to chryfalis the 11th of July, the moth appeared July 31; from which it appears certain that he reared it from the caterpillar. He has not, however, given a figure of it in the Aurelian, or any other of his publications.

PLATE

164

PLATE CLXIV.

TENTHREDO ROSÆ.

HYMENOPTERA.

Wings four, generally membraneous. Tail of the females armed with a sting.

GENERIC CHARACTER.

Abdomen of equal thickneſs, and cloſely connected to the thorax. Sting, ſerrated, between two valves. Second wings ſhorteſt.

SPECIFIC CHARACTER

AND

SYNONYMS.

Antennæ, head, and thorax black, with a yellow ſpot on each ſide of the latter. Abdomen yellow. A black ſpot on the anterior margin of the wings.

TENTHREDO ROSÆ: antennis ſeptemnodiis nigra, abdomine flavo, alarum anteriorum coſta nigra.
Syſt. Ent. 322. 26.
Fab. Spec. 1. 413. 39.

Tenthredo Roſæ antennis clavato, filiformibus nigra abdomine flavo, alarum anticarum coſta nigra.
Lin. Syſt. Nat. 2. 925. 30.
Fn. Sv. 1555.

Tenthredo crocea thorace ſupra, capite alarumque margine exteriori nigris. *Geoff. Inſ.* 2. 272. 4.

I   *Tenthredo*

Tenthredo flava, antennis clavatis triarticulatis, capite thoraceque nigris, alis anticis nigro maculatis. *Degeer. Inf.* 2. 2. 279. 28. *tab.* 39. *fig.* 27.

Merian. Europ. tab. 144.

Goed. Inf. 2. *tab.* 3.

Scop. carn. 722.

Reaum. Inf. 5. *tab.* 14. *fig.* 10. 12.

In the larva ftate, this fpecies feeds on the leaves of the Rofe, and from that peculiarity it has received its fignificant fpecific name, *rofæ*. The larva cafts its fkin feveral times before it becomes a pupa, its exuviæ we frequently find adhering to rofe-leaves. When the larva is in its laft fkin it is yellowifh, inclining to orange, with many minute black fpecks, difpofed in ringlets, on every joint; but in the early ftages of its growth we find them of feveral fhades of colours, between green and orange, and fome partake of both colours, and are fpeckled with black, as in the laft fkin. The larva is very tender, and, we fufpect, is liable to fome diftemper of a very different kind from any noticed to affect other infects; it then appears fickly, and is covered with a whitifh down, or powder, which flies off on the flighteft touch. We have often found the larva of another fpecies of the fame genus covered with this kind of white powder, but as they always died, it is impoffible to determine to what infect they belonged.

In the pupa ftate, the outer cafe is not perfectly oval, but rather flattened on the fides; it is generally faftened on a ftalk. The perfect infect is found in great plenty during moft of the fummer months.

Several early fyftematic writers placed this infect amongft thofe whofe antennæ confifted of feven joints, or articulations: Whence *Linnæus* * included the number of the joints with the fpecific cha-

___

* In the laft edition of the *Syft. Nat.* " antennis feptemnodiis, &c." is changed for " antennis clavato filiformibus, &c."

racter;

racter ; and in the Species Infectorum *Fabricius* has followed the
fame arrangement.   Though with the affiftance of a microfcope we
may difcover in this, and other fpecies, the exact number of the
articulations defcribed, yet they are too minute to ferve as part of a good
fpecific character, which fhould, if poffible, be felected from the
moft confpicuous and peculiar parts of the infect.   Fabricius feems
to have been aware of this in his laft work, Entomologia Syftematica * ;
and has made a very judicious alteration ; though it appears fingular
for a fyftematic writer to change " Antennis filiformibus articulis.
7—9." for " Antennis inarticulatis, extrorfum craffioribus."

---

* *Tom.* 2. *p.* 109. 18.

165

PLATE CLXV.

PHALÆNA OXYACANTHÆ.

EALING's GLORY.

LEPIDOPTERA.

GENERIC CHARACTER.

Antennæ taper from the bafe. Wings in general deflexed when at reft. Fly by night.

Noctua.

SPECIFIC CHARACTER

AND

SYNONYMS.

Firft wings, dark brown, with two large irregular fpots of white and reddifh colour, and a broad fpace of the fame next the exterior margin: in feveral parts a fpeckling of fine blueifh green. Second wings, and body, plain brown.

PHALÆNA OXYACANTHÆ: criftata alis deflexis bimaculatis : margine tenuiori coerulefcente ; lunula alba.

Lin. Syft. Nat. 2. 852. 65.—*Fn. Sv.* 1207.

Fab. Spec. Inf. 2. 232. 114.—*Ent. Syft. Tom.* 3. *pars.* 2. *p.* 93. 277.

Wien. Verz. 70. 3.

Roef. Inf. 1. *phal.* 2. *tab.* 33.

Wilks. pap. 12. *tab.* 1. *c.* 1.

The

The caterpillar of this fpecies is found on the White Thorn, in April ; in May it becomes a pupa : the moth does not appear before September.

It will be readily conjectured, from its Englifh name, to be more frequently taken about *Ealing*, in *Middlefex*, than elfewhere, though it is not peculiar, like fome infects, to one place only. The cater-pillar is fmooth, or without any hairs ; it eats ravenoufly, is very fluggifh, and forms a fine filky web, in the ground, in which it paffes to the pupa ftate *. We find the moth very liable to variation in colours ; in fome fpecimens the green is very brilliant, in others the red ; and again, in others, the lunar white marks are very con-fpicuous. In fome fpecimens, natives of warm countries, we have feen them finer coloured than thofe from the northern parts of Europe.

---

* In the plate of this fpecies in Roeel's German Infects, the filky cone of the pupa is drawn in the convex part of a leaf.

166

PLATE CLXVI.

LIBELLULA GRANDIS.

LARGEST DRAGON FLY.

NEUROPTERA.

Wings four, naked, tranſparent, reticulated with veins, or nerves. Tail without a ſting.

GENERIC CHARACTER.

Mouth always armed with more than two jaws. Antennæ ſhorter than the thorax. Wings expanded. Tail of the male forked.

SPECIFIC CHARACTER

AND

SYNONYMS.

Thorax brown, with two oblique lines of yellow on each ſide. Abdomen red-brown, with white ſpots. Wings with a marginal ſpot.

LIBELLULA GRANDIS: alis glauceſcentibus, thoracis lineis qua-tuor flavis. *Lyn. Syſt. Nat.* 2. 903. 9.
Fn. Sv. 1467.

AESHANA GRANDIS: thorace lineis quatuor flavis, corpore varie-gato. *Fab. Syſt. Ent.* 424. 2.—*Spec. Inſ.* 2. p. 525. 133. 2.—*Ent. Syſt. T.* 2. p. 384. 2.

Libellula fulva, alis flaveſcentibus, thoracis lateribus lineis duabus flavis, fronta flaveſcente, cauda diphylla. *Geoff. Inſ.* 2. 227. 12.

I   *Libellula*

Libellula fufca, capite rotundato, thorace lineolis quatuor tranfverfis
luteis, alis flavicantibus, abdomine cylindrico.
Degeer. Inf. 2. 2. 45. *tab.* 20. *fig.* 6.
Libellula maxima vulgatiffima, alis argenteis.   *Raj. Inf.* 48. 1.
Roef. Inf. 2. *Aqu.* 2. *tab.* 2. *fig.* 1. 2?
Schœff. Icon. tab. 2. *fig.* 4.
Act. Nidros. 3. 412. *tab.* 6. *fig.* 9.

If we except a very fmall number of exotic *Libellulæ*, *L. Grandis*
is the largeft infect of the genus known : it is certainly the largeft
of the European fpecies.

It is not uncommon in woods ; but never flies far from the water.
In the larva ftate it lives in the water, and, like others of the fame
genus already defcribed in this work, does not quit it till it becomes
a winged creature.   In the larva ftate it alfo refembles in its manners
thofe voracious infects that devour fmaller infects, and in the winged
ftate it takes moths and other weak infects in its flight.   Is found in
moft of the fummer months.

P L A T E

167

PLATE CLXVII.

STAPHYLINUS RIPARIUS.

BANK ROVE-BEETLE.

COLEOPTERA.

GENERIC CHARACTER.

Antennæ moniliform*. Elytra not more than half the length of the abdomen. Wings concealed. Tail armed with two oblong vesicles.

SPECIFIC CHARACTER

AND

SYNONYMS.

Red-brown. Shells blue. Head and end of the abdomen black.

STAPHYLINUS RIPARIUS: *Lin. Syst. Nat. n.* 8. *p.* 684. *Ed.* 13.
  n. 9. *p.* 2038.—*Fn. Sv. n.* 846.
Staphylinus gregarius. *Scop. Carn. n.* 308. *ic.* 308.
Staphylin de rivages. *Degeer. Inf.* 4. *p.* 28. *n.* 14. *tab.* 1. *fig.* 18.
  Geoffr. Inf. 1. *n.* 21. *p.* 369.
  Paykull. monogr. Staphyl. *n.* 19. *p.* 27.
  Schäff. Icon. Inf. Ratisb. tab. 71. *fig.* 3.
  Harrer Beschr. d. Schäff. Inf. 1. *Th. n.*
  417. *p.* 253.
PÆDERUS RIPARIUS: rufus elytris coeruleis, capite abdominifque
  apice nigris. *Fab. Syst. Ent.* 1. *p.* 168.—
  Spec. Inf. T. 1. *p.* 339.—*Mant. Inf.* 1. *p.*
  223.—*Ent. Syst.* 2. *p.* 536.
Der Strandttraubenkäfer. Der Uferraubkäfer. *Panz. Faun. Inf.*
  Germ. Inhalt des neunten Hefts. tab. 11.

---

* Like a necklace of small beads.

I 2

The

The Staphylini were formerly known among Englifh collectors by the general appellation *Rove-Beetles;* we have in the prefent inftance adopted this Englifh name, and added the only fpecific diftinction which occurs likely to convey the meaning of Linnæus, when he named it *Riparius.*

All the infects of this genus are very voracious. The larvæ fo much refemble the perfect infects, that they can hardly be diftinguifhed from them. *Staphylinus Riparius* is found in moft parts of Europe. It frequents moift fandy places, and the fides of banks. Found in May. The natural fize and magnified appearance is given in the annexed plate.

P L A T E

168

PLATE CLXVIII.

PHALÆNA SATELLITIA.

SATELLITE MOTH.

LEPIDOPTERA.

GENERIC CHARACTER.

Antennæ taper from the bafe. Wings in general deflexed when at reft. Fly by night.

Noctua.

SPECIFIC CHARACTER

AND

SYNONYMS.

Firft wings, exterior margin indented : reddifh brown with feveral dark ftreaks acrofs : in the center a yellow fpot between two fmaller white fpots. Second wings greyifh.

PHALÆNA SATELLITIA : criftata alis deflexis dentatis brunneis: anticis puncto flavo inter punctula duo alba. *Fab. Spec. Inf.* 2. 230. 104. *Lin. Syft. Nat.* 2. 855. 176. *Roef. Inf.* 3. *tab.* 50.

The caterpillar of this moth feeds on whitethorn, currant and goofeberry-trees, &c. The chryfalis or pupa is enclofed in a ftrong web of a greyifh colour; it is of a dark brown colour. The caterpillar is found in June. In July or Auguft, the moth comes forth.

The upper wings of this moth have a very ftriking characteriftic; that is, the yellowifh lunar mark within two fmall fpots: from this character it has been aptly named Satelliti; and in Englifh, the Satellite Moth.

I 3

PLATE

169

PLATE CLXIX.

PAPILIO CARDAMINES.

ORANGE-TIP BUTTERFLY, or,

WOOD LADY.

LEPIDOPTERA.

GENERIC CHARACTER.

Antennæ terminate in a club.   Wings erect when at reft.   Fly in day-time.

* * * * * * Danai Candidi.

SPECIFIC CHARACTER

AND

SYNONYMS.

Wings rounded, edges very flightly fcalloped.   Above white, exterior half of the upper wings orange ; with a black fpot in the centre. Underfide of under wings marbled with green.   Female has no orange tip.

PAPILLIO CARDAMINES: alis rotundatis integerrimus albis: pofticis fubtus viridi marmoratis.   *Lin. Syft. Nat.* 2. 761. 85.—*Fn. Sv.* 1039.

Papilio minor alis exterioribus albis macula infigni crocea fplendentibus, interioribus fuperne albis, fubtus viridi colore variegatis.   *Raj. Inf.* 115.
Roef. Inf. pap. 2. *tab.* 8.
Schæff. Icon. tab. 91. *fig.* 1. 3.
————*Elem. tab.* 94. *fig.* 8.

PAPILIO CARDAMINES. *Fab. Spec. Inf. 2. 43. 179.*
Hafn. Icon. tab. 9. fig. 1.
Efp. pap. v. tab. 4. fig. 1.
——————— *tab. 27. fig. 2.*
Wilk. pap. 2. p. 50. tab. a. 5.
Robert. Icon. tab. 21.

Lady of the Woods. *Harris.*

———————————————

This pretty Butterfly may be taken in great abundance in the month of May. The caterpillar is found on various kinds of grafs and low herbage: Harris fays it feeds on Wild Cole; and other writers mention, Thlafpi Burfa Paftoris*, and Cardamine Pra-tenfis†.

The male infect only, has the bright orange colour on the wings, the female is white, with fome few marks of black: the underfide is beautifully marbled and mottled with green in both fexes.

The Caterpillar is common in May and June, and a later brood is found in July; about the latter end of which month it becomes a chryfalis: In May following the Butterfly is produced.

———————————————

* *Shepherd's Purfe.*   † Common Lady's Smock, or Cuckow-flower.

PLATE

170

PLATE CLXIX.

PHALÆNA SAMBUCARIA.

Swallow-tail Moth.

Lepidoptera.

GENERIC CHARACTER.

Antennæ taper from the bafe.   Wings in general deflexed when at reft.   Fly by night.

Geometra.

SPECIFIC CHARACTER

AND

SYNONYMS.

Wings angulated, pale yellow, with two tranfverfe lines on each. Second wings with a tail each, and two black fpots.

PHALÆNA SAMBUCARIA: pectinicornis, alis caudato angulatis flavefcentibus, ftrigis duabus obfcurioribus, pofticis apice bipunctatis. *Lin. Syft. Nat.* 2. 860. 203.—*Fn. Sv.* 122.

Phalæna feticornis fpirilinguis, alis patentibus fulphureis, linea duplici tranfverfa obfcuriori, inferioribus caudatis. *Geoff. Inf.* 2. 138. 58.

Phalæna

Phalæna media ochroleucos, alis ampliſſimis, exterioribus duabus lineis tranſverſis, e fulvo virentibus, interioribus, una diviſis.   *Raj. Inſ.* 177. 1.

Phalæna antennis filiformibus, alis latis angulatis luteis, ſtrigis duabus obſcurioribus.   *Degeer Inſ. Verſ. Germ.*
2. 1. 327. 3.

Albin Inſ. tab. 94.
Roeſ. Inſ. 1 *phal.* 3. *tab.* 6.
Petiv. Gazoph. tab. 51. *fig.* 6.
Wilks pap. 38. *tab.* 1. 6. 2.
Clerk. Icon. tab. 50. *fig.* 2.
Schœff. Icon. tab. 93. *fig.* 8.
Sepp. Inſ. 6. 1. *tab.* 1.
Wien Verz. 103. 1.

Inſects, when in the larva ſtate, have various means of protecting, or concealing themſelves from other ſpecies that would annoy them, as well as from birds who prey on them. This remark is partly juſtified by the ſubject of our annexed plate, the larva of which we find is not furniſhed with any means of defence when attacked: nor of agility to run away, or ſecrete itſelf from its enemies; but to compenſate for this, nature has formed it with a ſkin of ſuch a colour, and ſtructure, that its greateſt ſafety is in its inaction. We frequently ſee it faſtened by its hind feet to a ſmall twig or branch in ſuch a poſture, that unleſs it moves, it is ſcarcely poſſible to diſcover it. It is ſometimes in an erect poſition, at others with its head downwards, but in an oblique poſition; and, as it hangs in this manner, without the leaſt appearance of life for a conſiderable time, it exactly reſembles a ſmall twig of the branch to which it is attached.

The

PLATE CLXIX. 87

The caterpillars are not uncommon in April, or early in May. It feeds on feveral plants; particularly, when in confinement, it prefers bramble, or white thorn. It is found in the winged ftate in June, fo that it remains a very fhort time in chryfalis.

PLATE

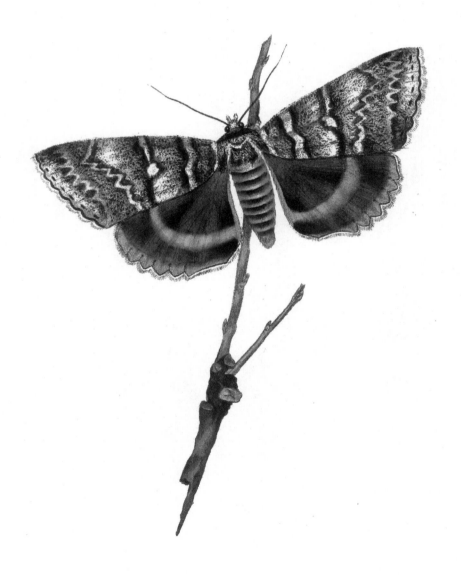

171

PLATE CLXX.

PHALÆNA FRAXINI.

CLIFDEN NON-PAREIL.

LEPIDOPTERA.

GENERIC CHARACTER.

Antennæ taper from the bafe. Wings in general deflexed when at reft. Fly by night.

Noctua.

Wings fcalloped, grey, with tranfverfe undulated bands of black ; in the centre of the wing ; fecond wings black, with a broad curved band of blue acrofs the middle.

PHALÆNA FRAXINI: criftata, alis dentatis cinereo nebulofis : pofticis fupra nigris : fafcia cærulefcente. *Lin. Syft. Nat.* 2. 843. 125.
> *Fn. Sv.* 1165.
> *Fab. Syft. Ent.* 602. 51.—*Spec. Inf.* 2. 221. 72.—*Ent. Syft. Nat. T.* 3. *p.* 2. 55. 152.

Phalæna feticornis fpirilinguis, alis deflexis, fuperioribus cinereo fufcoque, undulatis, inferioribus nigris, fafcia tranfverfa cœrulea. *Geof. Inf.* 2. 151. 83.

> *Roef. Inf.* 4. *tab.* 28. *fig.* 1.
> *Merian Europ. tab.* 46.
> *Ammir. Inf. tab.* 25.
> *Wilk. pap.* 45. *tab.* 1. *a.* 2.
> *Fyeſt. Arch. tab.* 15. *fig.* 1. 2.
> *Wien. Verz.* 90. 2.

From

From the Englifh name given to this beautiful and extremely fcarce moth, we learn that it has been taken at Clifden : we have alfo heard of its being found in other parts of England ; and, if we can rely on our information, a fpecimen was taken in July, 1795, in the fields.

We have never underftood that the larva had been found in this country. Feeds on the afh tree.

P L A T E

172

J^b L'Admiral Junr
Amsterdam

PLATE CLXXI.

THE

CATTERPILLAR

AND

CHRYSALIS

OF

PHALÆNA FRAXINI.

The rarity of this fubject muft plead our apology for the liberty we have taken in introducing it into our work. It is the only in- ftance in which we have given place to a copy from the works of others of any fubject, however rare. We have in our poffeffion a preferved fpecimen of the caterpillar of Phalæna Fraxini, fent from Germany; but as it is of that kind in which the colours and form cannot be preferved well, we have preferred giving an exact copy of the caterpillar as well as chryfalis, from the works of a refpecta- ble, but little known author, *Ammiral.* This author appears to have been fingularly fortunate in prefenting a figure of the caterpillar, when the accurate Roefel did not publifh a figure of the moth till his fourth volume, and was not then in poffeffion of the caterpillar.

Some of our readers will be perhaps furprifed to find that our figures precifely agree with thofe contained in the Aurelian of our countryman *Harris*; but whoever poffeffes the plates of *Ammiral,* will find that in the moft minute parts of Harris's plates, he has only traced and reverfed the originals of *Ammiral* throughout; and in many inftances by a clumfy imitation, in reverfing the foliage and flies, has even fpoilt the effect, and loft fight of the accuracy of them.

PLATE

173

PLATE CLXXII.

PAPILIO VIRGAUREÆ.

SCARCE COPPER BUTTERFLY,

LEPIDOPTERA.

Papilio ruralis. *Lin.*
Hefperia ruralis. *Fab.*

GENERIC CHARACTER.

Antennæ terminated in a club. Wings, when at reft, erect,
Fly by day.

SPECIFIC CHARACTER

AND

SYNONYMS.

Wings angulated. Upperfide of a fine bronze, or red copper
colour, with a black margin. Underfide light brown, with feveral
white fpots, fome having a black fpeck near the middle.

PAPILIO RURALIS VIRGAUREÆ. *Lin. Syft. Nat. n.* 253. *p.* 793.
edit. 12. *n.* 253. *p.* 2359.—*Faun. Suec.*
n. 1079.

Papilio ruralis Virgaureæ. Fab. Syft. Ent. n. 569. *p.* 126.—
Spec. Inf. 2. 569. *p.* 126.—*Mant. Inf.*
2. 721. *p.* 79.

K   *Hefperia*

Hesperia ruralis Virgaureæ: alis subangulatis fulvis: margine atro, subtus punctis, nigris albisque.

Fab. Ent. Syst. 4. 173. *p.* 309.

Le Bronzè.

Geoffr. Inf. 2. 35. *p.* 65.
Papil. d'Europ. tab. 44. *n.* 92.
Esper eur. Schmett. 1. *Th. tab.* 19. *fig.* 2.
Borkhausen eur. Schmett. 1. *Th.* 1. *p.* 141.
et p. 269.
Syst. Verz. d. W. Schmett. 1. *p.* 80.

L'Argus satiné.   *Ernst.*

Der Goldrathenfalter.   Der Feverpapilion.   *Panz. Faun. Inf. Germ.*

─────────────────────

A specimen of this very superb and rare butterfly has been taken at Cambridge. It has always had a place in the cabinets of English collectors of consequence; but we cannot learn by whom it was first discovered in this country. Papilia Virgaureæ and Papilio Hippothoe, has been frequently confounded with each other; but on a comparison, a material difference will be discovered.

Harris has made one error, which it is of importance to the English collector to correct; he says, " *Papilio Virgaureæ*, copper, feeds on grafs, found in June and August in meadows, is shining copper, spotted with black." From this it appears he could mean no other than the common copper butterfly, which is found in June and August in meadows, *Papilio Phlæas*; for though the scarce copper butterfly was probably found in his time, it must have been very rare;

PLATE CLXXII.   95

rare ; and he would not have omitted in his catalogue of Englifh Lepidopteræ, to mention an infect fo common as *Papilio Phlæas*, if he had noticed the other. He has alfo the fame error in his Aurelian.

PLATE

L

PLATE CLXXIV.

BUPRESTIS VIRIDIS.

GREEN BUPRESTIS.

COLEOPTERA.

GENERIC CHARACTER.

Antennæ fetaceous, and as long as the Thorax. Head drawn within the Thorax.

SPECIFIC CHARACTER

AND

SYNONYMS.

Linear, fhining blue and green in fhades, a few exceedingly minute fpots fprinkled over fome parts.

BUPRESTIS VIRIDIS: elytris integerrimis linearibus punctatus, corpore viridi elongato. *Fab. Spec. Inf.* 1. 281. 54.—*Syft. Ent.* 223. 38.—*Linn. Syft. Nat.* 2. 663. 25.—*Fn. Sv.* 762.

Bupreftis viridis nitida, corpore elongato, elytris linearibus fcabris integerrimus. *Degeer. Inf.* 4. 1. 33. 6. *tab.* 5. *fig.* 1.

Cucuius viridi cupreus oblongus. *Geoff. Inf.* 1. 127. 5.

Mordella ferraticornis. *Scop. Carn.* 190.

The Larva of this Infect feeds on the Birch-tree (Betula Alba). It is rarely met with in England ; and if we may form any opinion from the filence of Naturalifts, it is not common in any part of Europe.

F. I. Natural fize.

L   PLATE

PLATE CLXXV.

PHALÆNA SCHÆFFERELLA.

LEPIDOPTERA.

GENERIC CHARACTER.

Antennæ taper from the bafe. Wings in general deflexed when at reft. Fly by night.

TINEA.

SPECIFIC CHARACTER

AND

SYNONYMS.

Firft wings orange, with fpots and ftripes of filver: a deep black fringe. Second wings pale black.

TINEA SCHÆFFERELLA: *Linn. Syft. Nat.* 2. 898. 443.—*Fn. Sv.* 1409.

TINEA SCHÆFFERELLA: aliis nigris: difco flavo; ftrigis lineis duabus punctifque tribus argenteis. *Fab. Spec. Inf.* 2. 303. 79.—*Ent. Syft. Tom.* 3. *p.* 2. 322. 155.

Linnæus fays this beautiful little Infect feeds on the Chefnut. We found it on the Tanfey, in May, 1796.

It has not been figured by any author that has fallen under our infpection ; and the Synonyms given by Fabricius, in his laft work, refer only to the defcription given by Linnæus, and *Wien. Verz.* * 138. 21.—The fpecific name was adopted by Linnæus, and continued

---

* Catalogue of Infects found near Vienna.

L 2

by

by Fabricius, in honour of Schæffer, Author of the Infecta Ra-
tifbonenfia, and Fundamenta Entomologica. *Quarto.* 1747.

The natural fize of this Infect is given at the bottom of the
Plate ; and as it was too minute to admit of the elegant colouring
of the original, two figures of its magnified appearance is given
above, one in its refting pofition, the other with the Wings expanded.

PLATE

176

PLATE CLXXVI.

NOTONECTA STRIATA.

STRIATED BOAT FLY.

HEMIPTERA.

Upper Wings femi-cruftaceous, not divided by a ftraight future, but incumbent on each other. Beak curved downward.

GENERIC CHARACTER.

Beak inflected. Antennæ fhorter than the Thorax. Wings croffed. Hind Feet hairy, and formed for fwimming.

SPECIFIC CHARACTER

AND

SYNONYMS.

Depreffed. Head and Legs yellow; reft pale brown, with numerous minute fpots and ftreaks of dark brown.

NOTONECTA STRIATA. *Lin. Syft. Nat.* 2. 712. 2.—*Sv.* 904.
SIGARA STRIATA: elytris pallidis: lineolis tranfverfis undulatis numerofiffimis fufcis. *Fab. Spec. Inf.—Ent. Syft. T.* 4. 207. 2.
Corixa. *Geoffr. Inf.* 1. 478. 1. *tab.* 9. *fig.* 7.
Stoll. Cicad. 2. *tab.* 15. *fig.* 13. *B.*
Roef. Inf. 3. *tab.* 29.
Schæff. Elem. tab. 50.
———— *Icon. tab.* 97. *fig.* 2.
Fyef. Helvet. 25. 469.

There

There are two varieties of this ſpecies: one kind being at leaſt twice the ſize of the other; in every other reſpeét they perfeétly agree. This Inſeét is commonly ſeen on ſtill waters, in the Summer; when they cauſe a gentle agitation of the ſurface, by the quickneſs of their motions, and row along on their back, with their hinder legs, which are formed for ſwimming. Both kinds are ſhown in the Plate, Fɪɢ. I. and II. Fɪɢ. III. is the largeſt ſort magnified to exhibit the curious markings of the Elytra.

P L A T E

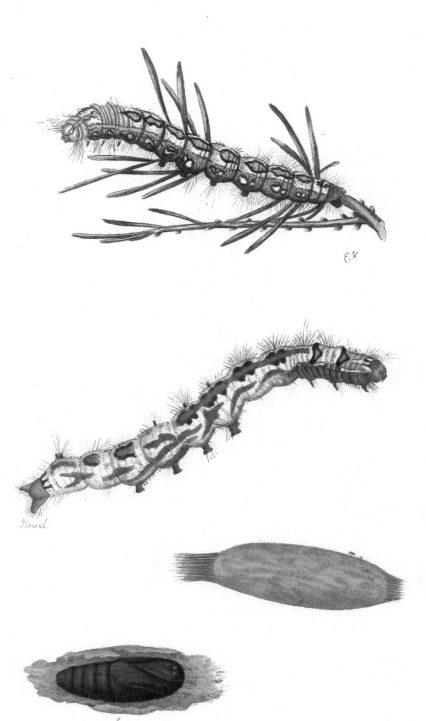

177

PLATE CLXXVII.

THE

CATERPILLAR

OF

PHALÆNA PINI.

We have introduced in the annexed plate, figures of the Caterpillars of Phalæna Pini, copied from the works of the two moſt
accurate entomologiſts that have deſcribed or figured the inſects of
any part of the European continent; and though unfortunately the
deſcriptions are written in a language ſo little underſtood as to be
wholly uſeleſs; the figures are very intereſting. In this inſtance we
have deviated no more from our original plan than when we introduced the larva of *Sphinx Euphorbiæ,* and *Phalæna Fraxini*; and we
flatter ourſelves in thus endeavouring to give the hiſtory of a rare
inſect complete, the approbation expreſſed by our ſubſcribers, on
former occaſions, will not be withheld on the preſent.

Roeſel, in 1746, publiſhed the *Inſecten Beluſtigung*; in which
work we find a figure of the Caterpillar of Phalæna Pini: it accords
with the deſcription given by Fabricius; perhaps the deſcription
was taken from Roeſel's plate. " Larva ſubcaudata, albo griſeo
fuſcoque variegata, collaribus coeruleis: punctis utrinque rufis."
Fab. Syſt. T. 3. *p.* 2. 426. 62.

Kleeman, the relation and ſucceſſor of Roeſel, in the third part of
his ſupplement, Plate 6. fig. 7*. has ſhewn the Caterpillar of this

---

* Publiſhed in 1793.

L 4

inſect

infect in another fkin, or probably it is the Caterpillar of the male, Roefel having only the female in his works; in this fpecimen the colours are bright, and it is particularly diftinguifhed by the collar being red inftead of blue.—As this part of his work is fcarcely known, and has not yet been noticed by *Fabricius*, we cannot collect the opinion of any fyftematical writer, whether it be the other fex, or only a different fkin.

The pupa we received with the moths; and the eggs figured in plate 178, were taken from the body of the female.

P L A T E

178

PLATE CLXXVIII.

PHALÆNA PINI.

PINE LAPPET MOTH.

LEPIDOPTERA.

GENERIC CHARACTER.

Antennæ taper from the bafe. Wings in general deflexed when at reft. Fly by night.

Bombyx.

Antennæ of the male feathered.

SPECIFIC CHARACTER

AND

SYNONYMS.

Firft wings grey, fpeckled with brown: a broad fpace of red brown acrofs each, and a triangular white fpot near the anterior margin.

BOMBYX PINI: alis reverfis grifeis: fafcia ferruginea punctoque
    triangulari albo. *Linn. Syft. Nat.* 2. 814. 24.—*Fn. Sv.*
    1104.—*Fab. Syft. Ent.* 3. *p.* 2. 426. 62.
    Merian. Europ. tab. 22.
    Wilks. pap. 29. *tab.* 3. *b.* 5.
    Roef. Inf. 1. *phal.* 2. *tab.* 59.
    Schæff. Icon. tab. 86. *fig.* 1—3.
    Kleman. Inf. 2. *Suppl. pl.* 6. *fig.* 7.

The Pine Lappet Moth is one of thofe fpecies of infects, that we can have no doubt are natives of this country, from the concurrent
             teftimony

teſtimony of the reſpectable authors; though from the ſcarcity of many amongſt them, we ſhould be ſcarcely inclined to admit them into an Engliſh collection without ſuch authority. Perhaps the rarity of ſome of thoſe inſects ſhould be rather attributed to the little attention beſtowed on the ſcience of Entomology by ſuch as reſide in parts of the kingdom that are moſt favourable to the increaſe of inſects in general; or to thoſe particularly rare ſpecies that are local, or feed only on plants of one kind; ſuch as the *Sphinx Euphorbiæ*, and many others.

Wilks has given the Pine Lappet Moth in the third plate of the Engliſh butterflies. *Harris* has not figured it in the Aurelian *, but in the Pocket Companion † he not only deſcribes it amongſt the Engliſh Lepidoptera, but ſays, the time of its changing into Chryſalis is *May*, and that it appears in the winged ſtate in June; from this we muſt ſuppoſe, that he had reared it from the Caterpillar. *Berkenhout*, in his ſynopſis of the natural hiſtory of Great Britain ‡, has given it without heſitation as an Engliſh inſect; and the authority of a little tract on inſects, by Martin §, may be adduced as a further confirmation of its being a native of this country.

This Inſect is not uncommon in Germany. Schæffer has figured it amongſt the inſects that are to be found in the environs of Ratiſbon; and Roeſel, without conſidering it a local ſpecies, has given it as a native of Germany. Whether it is found in other parts of Europe, except Switzerland and Germany, we cannot decide; but we have the preciſe ſpecies from *Georgia* in *North America*.

We obſerve a conſiderable difference between the colouring of this moth in the works of *Schæffer* and *Roeſel*, which is the more remarkable, as they both deſcribe the inſects of the ſame country; the figure given by the latter is much darker in the cheſnut colour, and the grey has no appearance of an intermixture of red ſpecks and markings, like that figured in *Schæffer*, which inclines very much to red or fleſh colour throughout. Roeſel has only figured the female; Schæffer has given both ſexes.

---

* Publiſhed in 1766.   † 1776.   ‡ 1789.   § 1785.

PLATE

PLATE CLXXIX.

PHALÆNA OO.

HEART MOTH.

LEPIDOPTERA.

GENERIC CHARACTER.

Antennæ taper from the base. Wings in general deflexed when at rest. Fly by night.

NOCTUA.

Antennæ like a bristle.

SPECIFIC CHARACTER

AND

SYNONYMS.

Wings buff, streaked, and marked with red-brown: and a double o in the middle of upper wings.

NOCTUA Oo: cristata alis deflexis cinerascentibus ferrugineo strigosis oo notatis. *Lin. Syst. Nat.* 2. 832. 81.—*Fn. Sv.* 1139.
    Fab. Syst. Ent. t. 3. *p.* 2. 247.
    Wien. Verz. 87. 1.
    Roes. Ins. 1. *Phal.* 2. *tab.* 63.

This Moth is far from common. It is found on the oak, in the Caterpillar state, late in the summer; changes to chrysalis in the first
week

week of October; the fly appears late in April, or early in May.
Harris greatly miftook the meaning of Linnæus, when he fays,
" Linnæan name, *Sphinx* Oo."

P L A T E

PLATE CLXXX.

ASILUS CRABRONIFORMIS.

HORNET FLY.

DIPTERA.

Wings, two.

GENERIC CHARACTER.

Trunk horny? long ſtraight, bivalved.

SPECIFIC CHARACTER

AND

SYNONYMS.

Body hairy ; the three ſegments next the thorax black, the four others yellow.

ASILUS CRABRONIFORMIS : abdomine tomentoſo antice ſegmentis tribus nigris poſtice flavo inflexo. *Fab. Spec. Inſ.* 2. 461. 5.—*Linn. Syſt. Nat.* 2. 1007. 4.

Aſilus ferrugineus abdominis articulis prioribus atris, poſteribus quatuor flavis. *Geoff. Inſ.* 2. 468. 3. *tab.* 17. *fig.* 3.

Aſilus ſubhirſutus, antennis ſetigeris, abdomine antice nigro poſtice flavo fulvo. *Degeer. Inſ.* 6. 244. 7. *tab.* 14. *fig.* 3.

Muſca maxima crabroniformis. *Raj. Inſ.* 267.

Erax crabroniformis. Scop. carn. 974.

Schæffer. Icon. tab. 8. *fig.* 15.

——— *Elem. tab.* 13.

This

This is a very confined genus. Fabricius in the Species Infectorum describes only thirty-three kinds; of these not more than eight are natives of this country. The Asilus Crabroniformis is the largest, and is not uncommon in some places in the summer, particularly frequenting wet meadows, and flying busily about the middle of the day amongst flowers.

Its probofcis is a curious instrument; the sting of it is very painful, and causes a swelling.

P L A T E

LINNÆAN INDEX.

TO

VOL. V.

COLEOPTERA.

HEMIPTERA.

LEPIDOPTERA.

M

Papilio

INDEX.

NEUROPTERA.

HYMENOP-

I N D E X.

HYMENOPTERA.

DIPTERA.

APTERA.

ALPHA-

ALPHABETICAL INDEX

T O

V O L. V.

Oo,

I N D E X.

ERRATA.

ERRATA.

Page 85, *for* Plate 169, *read* Plate 170.

Page 89, ———— 170, ———— 171.

Page 90, line 5, *after* fields, *add* near Hoxton.

Page 91, *for* Plate 171, *read* Plate 172.

Page 93, ———— 172, ———— 173.